Electromagnetic Field Computation by Network Methods

Leopold B. Felsen · Mauro Mongiardo ·
Peter Russer

Electromagnetic Field Computation by Network Methods

Springer

Prof. Leopold B. Felsen
Boston University
Dept. Aerospace &
Mechanical Engineering
110 Cummington St.
Boston MA 02215
USA

Prof. Mauro Mongiardo
Università Perugia
Dipartimento di
Ingegneria Elettronica e
dell'Informatione
Via G. Duranti, 93
06125 Perugia
Italy
mauro.mongiardo@gmail.com

Prof. Dr. Peter Russer
TU München
Fak. Elektro- und
Informationstechnik
LS für Hochfrequenztechnik
Arcisstr. 21
80333 München
Germany
russer@tum.de

ISBN 978-3-642-10104-5 e-ISBN 978-3-540-93946-7

DOI 10.1007/978-3-540-93946-7

Cover design: eStudio Calamar S.L.

Printed on acid-free paper

9 8 7 6 5 4 3 2 1

springer.com

One moment in annihilation's waste,
One moment, of the well of life to taste–
The stars are setting, and the caravan
Draws to the dawn of nothing–oh, make haste!

...

Since nothing remains to us from all that exists,
Since everything that exists is to be doomed, presumably
Whatever is permanent does not appear in the world,
And all established entities are inexistent.

Omar Khayyam

Preface

Electromagnetic field computations in either man-made or natural complex structures pose challenging problems with respect to electromagnetic wave propagation modeling, microwave circuit and antenna design, electromagnetic compatibility issues, high bit rate and ultra-wide band communications, biological hazards and numerous other problems. Since different problems exhibit specific combinations of geometrical features and scales, material properties and frequency ranges no single method is best suited for handling all possible cases: instead, a combination of methods is needed to attain the greatest flexibility and efficiency.

Naturally, with progress of computing facilities, the main focus has shifted from analytical computations to numerical ones. However, in many instances, the computations are performed in order to design a certain component, such as an antenna or a filter. Dealing with design and optimization problems requires not only the modeling of a given structure but also the evaluation of the sensitivities to parameter changes. In these cases it is worthwhile to attain the highest numerical efficiency in order to be competitive.

The present scenario witnesses the use of several different methods that, apart for a few noticeable exceptions, are not merged together. Clearly, from the efficiency point of view it would be desirable to solve the problem at hand in the most efficient way, thus subdividing the computational space in various subregions and by employing in each subregion the most satisfactory approach. Moreover, while the above procedure has been followed in several specific contributions, it is also important that the sought approach can be *systematically* employed for all cases. Of particular relevance are the rigorous treatment of the field at boundaries and the appropriate field representations inside bounded or unbounded regions.

The common ground which allows to achieve solutions that are rigorous, preserve energy conservation, and that can unify different methods, is the use of network theory, i.e. a rigorous translation of our field problem into an equivalent network problem. In particular the field at boundaries can be rigorously represented by using the Tellegen theorem for fields, which provides the generalized transformer network representation. In fact, a boundary can be seen as a region of zero volume in which no energy is stored neither dissipated, exactly as in a transformer. A region of finite volume, instead, when lossless can be seen as a resonator and its behavior may be described in terms of its resonances. Also, field propagation in an infinite region can be described in terms of spherical transmission lines, which provide an infinite, discrete, set of modes traveling along the radial direction. Such scenario, to our knowledge, has not been presented systematically in a book and, in our humble opinion, deserve instead some considerations. The aim of this book is therefore to illustrate with some detail how it is possible to describe whatever realistic electromagnetic field problem in terms of network elements, i.e. generalized transformers, RLC elements and transmission lines. The plan and content of the book is described in some detail in Chapter 1 and thus is not detailed here.

The reader may be interested in the genesis of this manuscript. Ties with Leopold Felsen were initiated through his invited attendance of the "International Workshop on Discrete Time Domain Modeling of Electromagnetic Fields and Networks", which convened in Munich in October 1991. Over a 14 years period we have had a fruitful scientific cooperation. It was 1996 and two of us, Leopold Felsen and Mauro Mongiardo were staying as Visiting Scientists with the third one, Peter Russer at the Technische Universität München. A topic that was often debated was that of complexity and how to find a systematic approach to compute electromagnetic field in complex structures. For those who have a more deep knowledge of the personality of Leopold Felsen it would not be difficult to believe that not every discussion was a smooth one. Nonetheless, after some time, we have found a considerable agreement on the procedure synthetically illustrated above. From this starting point we have worked on a sequence of papers illustrating the procedure, and in particular on triplet that was later published in a special issue of the International Journal for Numerical Computation. Also, a few other contributions increased our belief in this approach. A few years later, we agreed to start to work on a monograph on this subject and several other vivid discussions followed. Also a plan of the chapters started to evolve and after a certain time we initiated the actual work on the book. Our aim was to introduce the reader gradually with respect to the novelties; to this end we have started the book with standard electromagnetic theory.

As a large part of the book was already assembled and reviewed, the health conditions of Leopold Felsen deteriorated significantly leading to his untimely departure. This event left us with a deep sorrow for we greatly missed Leopold Felsen and his invaluable suggestions. The monograph project we tried to make what would have pleased him most. Since at that time Leopold Felsen has already contributed to the writing and corrections of the first four chapters of this book we decided to leave them in the form he was comfortable with. Accordingly, our task for this chapters has been only to implement his handmade corrections and improve figure qualities and other minor details. Also Chapter 5, although not yet complete, was already discussed with Leopold Felsen and agreed by him. The completion of this chapter and some other refinements have put the book in a condition that seems appropriate for disseminating the main ideas contained in it.

We are grateful to Leopold Felsen, for the instructive and pleasant time spent together. In Leopold Felsen we admired not only the exceptional scientist but also

a strong human character who has confronted his life's challenges with strength, courage and honesty and in the spirit of reconciliation.
We would like to express our appreciation to Patrizia Basili, Christos Christopoulos, Nikolaus Fichtner, Roberto Sorrentino and Cristiano Tomassoni for many helpful discussions. We thank Nikolaus Fichtner and Uwe Siart for support in solving typesetting problems. A particular thank goes to Christiane Wangerek who with her constant assistance has made possible for us to concentrate on the scientific part and rely on her superb organizational skills. We also would thank Leopold Felsen's son Michael Felsen who always has been very close to his father and has also taken the task of keeping us informed about his health and finally has encouraged us to finish this project. We would also like to express a sincere thank to our family members that have tolerated our secluded time and have provided constant and strong support to our effort.

Munich and Perugia,
November 2008

Mauro Mongiardo
Peter Russer

Contents

1

Introduction

I Motivation

Many applications in science and technology rely increasingly on electromagnetic field computations in either man-made or natural complex structures [1]. Wireless communication systems, for example, pose challenging problems with respect to field propagation prediction, microwave hardware design, compatibility issues, biological hazards, etc. Because different problems have their own combination of geometrical features and scales, frequency ranges, dielectric inhomogeneities, etc., no single method is best suited for handling all possible cases; instead, a combination of methods (hybridization) is needed to attain the greatest flexibility and efficiency.

The above considerations apply especially to the development of an "electromagnetic virtual laboratory" where experiments are simulated via computers. This type of virtual laboratory will probably become of increasing importance in the future for the analysis and design of electromagnetic structures. It is also noted that the availability of steadily increasing computing facilities has not lessened the need for efficient methods of electromagnetic field computation. This is readily understandable especially in the highly competitive design of microwave components. Success in this endeavor relies on more efficient techniques for electromagnetic field computation, which can be achieved by using hybrid techniques.

The necessity for hybrid methods has already been discussed in the past in overview papers by E.K. Miller [2] and G. A. Thiele [3]. Hybrid methods applied to scattering and antenna problems have been treated by L.N. Medgyesi-Mitschang and D.S. Wang [4–8], W.D. Burnside et al [9], G.A. Thiele and T.H. Newhouse [10], T.J. Kim and G.A. Thiele [11] and D.P. Bouche, F.A. Molinet and R. Mittra [12]. U. Jakobus and F.M. Landstorfer have devised techniques that combine the method of moments and the geometrical theory of diffraction or physical optics [13, 14]. Similarly, numerical methods such as finite elements (FEM) method or finite differences (FD) method, have been considered by D.J. Hoppe et al [1] and X. Yuan, et al [15] in conjunction with the method of moments (MoM). R. Khlifi and P. Russer developed a hybrid method combining the transmission line matrix (TLM) method and the time-domain MoM for the accurate modeling of the transient interference between remote complex objects [16,17]. D. Lukashevich et al have combined TLM method and the mode matching method providing an efficient tool for full-wave analysis of transmission lines and discontinuities in RF-MMICs [18, 19]. P. Lorenz and P. Russer have developed a TLM multipole expansion method for modeling of complex radiating structures [20–23]. The hybrid method presented by N. Fichtner, S. Wane, D. Bajon and P. Russer [24], combines the TLM and the TWF methods. Integral equations have been investigated by T. Cwik, V. Jamnejad and C. Zuffada [25, 26], and boundary integral methods have been developed by K. Ise and M. Koshiba [27]. Modal techniques have been treated by M. Mongiardo and R. Sorrentino [28] and G.C. Chinn et al [29]. Multipole methods have been used by N. Lu and J.M. Jin [30]. Combinations of boundary-contour and mode-matching methods have been investigated by J.M. Reiter and F. Arndt [31]. A hybrid electric field integral equation (EFIE) and magnetic field integral equation (MFIE) method for radiation and scattering problems referred to as the hybrid EFIE-MFIE (HEM) method has been proposed by R.E. Hodges and Y. Rahmat-Samii [32]. This list of contributions, though necessarily incomplete, indicates that this topic is of considerable interest. The methods listed above have typically been applied to solve a specific class of problems efficiently by matching the method to the perceived phenomenology, as in [9–11]. Despite these apparent diversities there are certain features common to all hybrid methods; namely, that the overall problem gets partitioned in a problem-matched manner.

In this book, we propose what we would like to call an *architecture*, i.e. a structure that addresses complexity systematically and with reasonable generality. We emphasize at the outset that an architecture does not *solve* a problem but it can provide a systematic framework for proper problem formulation. By that way such an architecture concept can contribute considerably to an efficient problem solution. In a three-part sequence of papers, L.B. Felsen, M. Mongiardo and P. Russer [33–35] already have outlined an architecture for a systematic an rigorous treatment of electromagnetic field representations in complex structures.

Our suggested architecture accommodates use of different numerical methods as well as alternative Green's function representations in each of the subdomains resulting from partitioning of the overall problem. The subdomains are characterized by *subdomain relations* and by *connection networks* between subdomains motivated by the problem topology. This is similar to what is customary in circuit theory and permits a phrasing of the solution of EM field problems in complex structures by network–oriented methods, which are also valuable from a numerical viewpoint. The classical problem of waveguide step discontinuities has been treated from the perspective of the *generalized network formulation* [36–40].

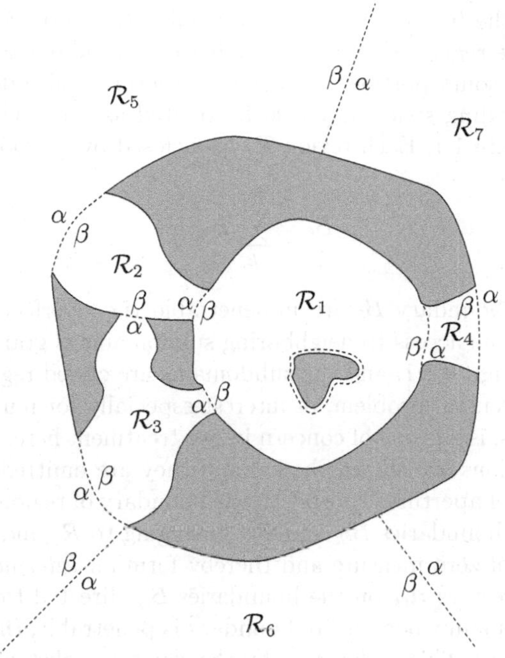

Fig. 1.1. Partitioning of the problem space into different regions denoted by $\mathcal{R}_\ell$ which are separated by boundaries $\mathcal{B}_{\ell k}$ (dashed curves); in this notation the first index refers to the region under consideration and the second index refers to the boundary with an adjacent region. The shaded regions are either perfect electric conductors (PEC) or perfect magnetic conductors (PMC).

II The Architecture

We assume that the problem geometry, the sources of excitation and the field observables (to be measured on desired reference surfaces) are specified. The architecture is based on three foundations

- Problem Partitioning
- Network Representations
- Methodological Hybridization.

II.1 Problem Partitioning

The principal task is the *partitioning* of the overall complex problem domain into subdomains which are selected so as to facilitate numerical and/or analytical treatment [35].
Consider a complex overall problem space $\mathcal{R}$ which is partitioned into $\mathcal{N}_R$ subdomains $\mathcal{R}_\ell$, $\ell = 1 \ldots \mathcal{N}_R$, (Figure 1.1). Any two subdomains $\mathcal{R}_\ell$ and $\mathcal{R}_k$ are

connected across the interface $\mathcal{B}_{\ell k}$, with the subscripts ordered so that the first index identifies the region of interest and the second index identifies an exterior region. Whenever some portion is an open structure embedded in unbounded space, this surrounding space may also be treated as a region, e.g. regions $\mathcal{R}_5$, $\mathcal{R}_6$ and $\mathcal{R}_7$ in Figure 1.1. Each region $\mathcal{R}_\ell$ is enclosed by the boundary

$$\mathcal{B}_\ell = \sum_{k}^{K_\ell} \mathcal{B}_{\ell k}\,. \tag{1.1}$$

When parts of a boundary $\mathcal{B}_\ell$ are impenetrable (i.e., perfect electric or magnetic conductors) the access to neighboring subdomains is granted via apertures (ports) $\mathcal{B}_{\ell k}$ as in Figure 1.1, and the subdomains are closed regions. This special case of the more general problem, of interest especially for multiport waveguide and cavity systems, is the one of concern in our treatment here. In Figure 1.1, the impenetrable portions are shown shaded and they are omitted from the sum in (1.1); the number of apertures (ports) on the boundary of region $\mathcal{R}_\ell$ is denoted by K_ℓ. Two adjacent boundaries $\mathcal{B}_{\ell k}$ and $\mathcal{B}_{k\ell}$ belonging to $\mathcal{R}_\ell$ and $\mathcal{R}_k$, respectively, enclose a volume of zero measure and thereby form an *interface*. We also introduce the normal vectors $\boldsymbol{n}_{\ell k}$ on the boundaries $\mathcal{B}_{\ell k}$ directed toward the exterior of $\mathcal{R}_\ell$. For a subdomain whose entire boundary is penetrable, the access "port" is that entire boundary. This is depicted by the separate "obstacle" in the interior of $\mathcal{R}$; for simplicity the obstacle shall be regarded as perfectly conducting but this restriction can readily be removed. On each boundary $\mathcal{B}_{\ell k}$, as seen from $\mathcal{R}_\ell$, we shall *specify primary exciting fields* and then *determine* via the corresponding Green's function representations the resulting *secondary* field response generated on $\mathcal{B}_{\ell k}$ through interaction with the interior of $\mathcal{R}_\ell$. The choice of primary and secondary fields affects the type of *boundary conditions* pertaining to a particular subdomain and, thereby, the corresponding *alternative Green's function* representation. To separate one subdomain from adjacent regions, we shall apply the equivalence theorem: for example if, on one aperture, *electric* fields are selected as the primary fields, then the *magnetic* fields are the secondary fields, yielding an admittance description with the aperture replaced by a perfect electric conductor. For each particular selection of primary and secondary fields, the corresponding convergence properties, wave patterns and wave phenomenology determine the problem strategy. Finally, a note concerning terminology; we have used the words primary and secondary field variables which can be substituted with the words independent and dependent field variables if preferred.

II.2 Network Representations

Next we distinguish between *subdomain relations* and *connection relations*. For each subdomain and the corresponding boundary conditions, the secondary fields express how that subdomain responds to the excitation provided by the primary field. Each subdomain with its *subdomain relations* is therefore distinct from the other subdomains. We note that the subdomains can either be of finite volume or infinite volume (i.e. extending up to infinity). We will see that for subdomains of finite volume always exists a Green's function representation in terms of resonant modes which leads, after discretization, to a rigorous network equivalent. In particular, if the subdomain exhibits a preferred waveguiding direction the network representation may also be expressed by using transmission lines. When the subdomain is of infinite volume, i.e. extends up to infinity, it is advantageous to enclose it in a spherical boundary and to use spherical transmission lines to represent wave propagation toward infinity. With this device also infinite regions find a rigorous network representation.

Connection networks implement the topological relationships for fields; i.e. the continuity of the tangential field components at common interfaces between adjacent subdomains. Models of partitioning for electromagnetic field computation correspond to network models used in circuit theory as follows:

- relations at boundaries between adjacent subdomains ↔ topological relations in a network;
- subdomain relations per se ↔ laws governing the behavior of circuit elements such as resistors, inductors, capacitors, etc.

The application of network–oriented methods to electromagnetic field problems can contribute significantly to the problem formulation and solution methodology. The field problem can be treated systematically by the segmentation technique and by specifying canonical Foster representations for the subcircuits. Connections between different subdomains are obtained by selecting the appropriate independent field quantities via Tellegen's theorem. For each subdomain, as well as for the entire circuit, an equivalent circuit extraction procedure is feasible, either in closed form for subdomains amenable to analytical description or via the relevant pole structure description when a numerical solution is available. Network concepts in electromagnetics allow the application of complexity reduction methods to the state equations describing the discretized electromagnetic field. The application of system identification and parameter estimation methods for reduction of computational time and automatic generation of lumped element equivalent circuits is also feasible.

II.3 Methodological Hybridization

For each subdomain we also have the option to *select a specific numerical method* best suited for its characterization. For example, we may divide the structure in such a way that it is convenient to use an integral equation approach in one subdomain, a finite element solution in another subdomain and a boundary element method in a third subdomain. We may denote this type of hybrid computation of the electromagnetic field as *external hybridization.*

Next we introduce, for each subdomain a strategy that can lead to efficient methods for the solution of the electromagnetic field equations. Based on any possible symmetries present in the subdomain, it is suggestive to advocate use of analytical solutions and, when necessary, couple these to numerical approaches such as finite elements or finite differences. We may refer to this type of hybridization as *internal hybridization* since it applies in the interior of each subdomain. Several well–known methods already embed the above strategy: for example, the method of lines solves the equations analytically in one direction, and use a finite difference grid in the other directions. We suggest the systematic exploitation of this approach, exploring different possibilities, such as combining finite element and modal techniques, or finite difference and finite element methods.

III Organization of the Book

The book is organized into five chapters. This first chapter contains the introduction and the motivations for this work.
The second chapter collects general material useful for reference purposes and introduce notations. After summarizing Maxwell's equations, we review general electromagnetic theorems and concepts and the use of field potentials to achieve scalarization of the vector field problems. The technique of variable separation is also reviewed for rectangular, spherical and circular cylindrical coordinate systems, followed by the a general discussion on the Sturm–Liouville problems.
In Chapter 3 we review the customary field expansions in waveguiding regions. We recall modal representations of fields and their sources and field representation in this particular type of subdomains is obtained in terms of transverse vector eigenfunctions (derived in turn from transverse scalar eigenfunctions) and longitudinal scalar voltages and currents.
Chapter 4 deals with simple two–dimensional examples, in rectangular and cylindrical coordinate systems, which elucidate the phenomenology of waveguide propagation and the application of the techniques introduced in the previous chapters.
Chapter 5 is the central part of this book, since it deals with the systematic rigorous network representation of electromagnetic field problems. We introduce the connection network for the regions of zero volume that constitutes the interfaces between different subdomains. Properties for the connection network are derived from the field version of Tellegen's theorem. Then we introduce the network representation available for subdomains of finite volume, either in terms of resonant modes and in terms of transmission lines. Finally, we deal with subdomains extending up to infinity, i.e. subdomains of infinite volume, for which the spherical mode expansion is introduced and the relative network representation is obtained. We conclude this chapter by showing a general procedure for solving electromagnetic field problems via the Tableau equations.

Since several symbols are used in this work, in the appendix, in order to alleviate the mnemonics effort, we have summarized their meanings and the equation where they have been first introduced.

References

[1] D. J. Hoppe, L. W. Epp, and J. F. Lee, "A hybrid symmetric FEM/MoM formulation applied to scattering by inhomogeneous bodies of revolution," *IEEE Trans. Antennas Propagat.*, vol. 42, pp. 798–805, Jun. 1994.

[2] E. K. Miller, "A selective survey of computational electromagnetics," *IEEE Trans.Antennas Propagat.*, vol. 36, pp. 1281–1305, Sep. 1988.

[3] G. A. Thiele, "Overview of selected hybrid methods in radiating systems analysis," *Proc. IEEE*, vol. 80, pp. 66–78, Jan. 1992.

[4] L. N. Medgyesi-Mitschang and D. S. Wang, "Review of hybrid methods on antenna theory," *Ann. Teleccomunicat.*, vol. 44, no. 9, 1989.

[5] ——, "Hybrid solution for scattering from perfectly conducting bodies of revolution," *IEEE Trans. Antennas Propagat.*, vol. AP-31, pp. 570–583, Jul. 1983.

[6] ——, "Hybrid solution from large bodies of revolution with material discontinuities and coatings," *IEEE Trans. Antennas Propagat.*, vol. 32, pp. 717–723, Jul. 1984.

[7] ——, "Hybrid solution for large impedance coated bodies of revolution," *IEEE Trans. Antennas Propagat.*, vol. 34, pp. 1319–1329, Nov. 1986.

[8] D. S. Wang, "Current-based hybrid analysis for surface wave effects on large scatterers," *IEEE Trans. Antennas Propagat.*, vol. 39, pp. 839–850, Jun. 1991.

[9] W. Burnside, C. Yu, and R. Marhefka, "A technique to combine the geometrical theory of diffraction and the moment method," *IEEE Trans. Antennas Propagat.*, vol. 23, no. 4, pp. 551–558, Jul 1975.

[10] G. Thiele and T. Newhouse, "A hybrid technique for combining moment methods with the geometrical theory of diffraction," *IEEE Trans. Antennas Propagat.*, vol. 23, no. 1, pp. 62–69, Jan 1975.

[11] T. J. Kim and G. A. Thiele, "A hybrid diffraction technique-general theory and applications," *IEEE Trans. Antennas Propagat.*, vol. 30, pp. 888–897, Sep. 1982.

[12] D. P. Bouche, F. A. Molinet, and R. Mittra, "Asymptotic and hybrid techniques in electromagnetic scattering," *Proc. IEEE*, vol. 81, pp. 1658–1684, Dec. 1993.

[13] U. Jakobus and F. M. Landstorfer, "Improved PO-MM hybrid formulation for scattering from three-dimensional perfectly conducting bodies of arbitrary shape," *IEEE Trans. Antennas Propagat.*, vol. 43, pp. 162–169, Feb. 1995.

[14] ——, "Improvment of the PO-MoM hybrid method by accounting for effects perfectly conducting wedges," *IEEE Trans. Antennas Propagat.*, vol. 43, pp. 1123–1129, Oct. 1995.

[15] X. Yuan, D. R. Lynch, and J. W. Strohbehn, "Coupling of finite element and the moment methods for electromagnetic scattering from inhomogeneous objects," *IEEE Trans. Antennas Propagat.*, vol. 38, pp. 386–393, Mar. 1990.

[16] R. Khlifi and P. Russer, "Hybrid space discretizing method - method of moments for the efficient analysis of transient interference," *IEEE Trans. Microwave Theory Techn.*, vol. 54, no. 12, pp. 4440–4447, Dec 2006.

[17] ——, "A novel efficient hybrid TLM/TDMoM method for numerical modeling of transient interference," in *Proceedings of the 22th Annual Review of Progress in Applied Computational Electromagnetics ACES 2006, Miami, FL, USA*, Mar. 2006, pp. 182–187.

[18] B. Broido, D. Lukashevich, and P. Russer, "Hybrid Method for Simulation of Passive Struktures in RF-MMICs," in *Topical Meeting on Silicon Monolithic Integrated Circuits in RF Systems, April 9–11, 2003, Garmisch, Germany*, Apr. 2003, pp. 182–185.

[19] D. Lukashevich, B. Broido, M. Pfost, and P. Russer, "The Hybrid TLM-MM approach for simulation of MMICs ," in *Proc. 33th European Microwave Conference, Munich*, October 2003, pp. 339–342.

[20] P. Lorenz and P. Russer, "Hybrid transmission line matrix - multipole expansion TLMME method," in *Fields, Networks, Methods, and Systems in Modern Electrodynamics*, P. Russer and M. Mongiardo, Eds. Berlin: Springer, 2004, pp. 157 –168.

[21] ——, "Hybrid transmission line matrix (TLM) and multipole expansion method for time-domain modeling of radiating structures," in *IEEE MTT-S International Microwave Symposium*, Jun. 2004, pp. 1037 – 1040.

[22] ——, "Discrete and modal source modeling with connection networks for the transmission line matrix (TLM) method," in *2007 IEEE MTT-S Int. Microwave Symp. Dig. June 4–8, Honolulu, USA*, jun 2007, pp. 1975–1978.

[23] ——, "Connection subnetworks for the transmission line matrix (tlm) method," in *Time Domain Methods in Electrodynamics*, P. Russer and U. Siart, Eds. Springer, 2008.

[24] N. Fichtner, S. Wane, D. Bajon, and P. Russer, "Interfacing the TLM and the TWF Method using a Diakoptics Approach," in *2008 IEEE MTT-S Int. Microwave Symp. Dig. Atlanta, USA*, jun 2008, pp. 57–60.

[25] C. Z. T. Cwik and V. Jamnejad, "Modeling three-dimensional scatterers using a coupled finite-element integral-equation formulation," *IEEE Trans. Antennas Propagat.*, vol. 44, no. 4, pp. 453–459, Apr. 1996.

[26] ——, "Modeling radiation with an efficient hybrid finite-element integral-equation waveguide mode-matching technique," *IEEE Trans. Antennas Propagat.*, vol. 45, no. 1, pp. 34–39, Jan. 1997.

[27] K. Ise and M. Koshiba, "Numerical analysis of h-plane waveguide junctions by combination of finite and boundary elements," *IEEE Trans.Microwave Theory Tech.*, vol. 36, pp. 1343–1351, Sep. 1988.

[28] M. Mongiardo and R. Sorrentino, "Efficient and versatile analysis of microwave structures by combined mode matching and finite difference methods," *IEEE Microwave Guided Wave Lett.*, vol. 3, no. 7, pp. 241–243, Aug. 1993.

[29] G. C. Chinn, L. W. Epp, and D. J. Hoppe, "A hybrid finite-element method for axis symmetric waveguide-fed horns," *IEEE Trans. Antennas Propagat.*, vol. 44, no. 3, pp. 280–285, Mar. 1996.

[30] N. Lu and J. M. Jin, "Application of fast multipole method to finite-element boundary-integral solution of scattering problems," *IEEE Trans. Antennas Propagat.*, vol. 44, no. 6, pp. 781–786, Jun. 1996.

[31] J. M. Reiter and F. Arndt, "Hybrid boundary-contour mode-matching analysis of arbitrarily shaped waveguide structures with symmetry of revolution," *IEEE Microwave and Guided Wave Letters*, vol. 6, pp. 369–371, Oct. 1996.

[32] R. E. Hodges and Y. Rahmat-Samii, "An iterative current-based hybrid method for complex structures," *IEEE Trans. Antennas Propagat.*, vol. 45, no. 2, pp. 265–276, Feb. 1997.

[33] L. B. Felsen, M. Mongiardo, and P. Russer, "Electromagnetic field representations and computations in complex structures I: complexity architecture and generalized network formulation," *International Journal of Numerical Modeling: Electronic Networks, Devices and Fields*, vol. 15, pp. 93–107, 2002.

[34] ——, "Electromagnetic field representations and computations in complex structures II: alternative Green's functions," *International Journal of Numerical Modeling: Electronic Networks, Devices and Fields*, vol. 15, pp. 109 125, 2002.

[35] P. Russer, M. Mongiardo, and L. B. Felsen, "Electromagnetic field representations and computations in complex structures III: network representations of the connection and subdomain circuits," *International Journal of Numerical Modeling: Electronic Networks, Devices and Fields*, vol. 15, pp. 127–145, 2002.

[36] M. Mongiardo, P. Russer, M. Dionigi, and L. B. Felsen, "Waveguide step discontinuities revisited by the generalized network formulation," *1998 Int. Microwave Symposium Digest, Baltimore*, vol. 2, pp. 917 - 920, Jun. 1998.

[37] M. Mongiardo, P. Russer, C. Tomassoni, and L. Felsen, "Analysis of N–furcation in elliptical waveguides via the generalized network formulation," *1999 Int. Microwave Symposium Digest, Anaheim*, pp. 27–30, Jun. 1999.

[38] ——, "Analysis of N–furcation in elliptical waveguides via the generalized network formulation," *IEEE Trans. Microwave Theory Techn.*, vol. 47, pp. 2473–2478, Dec. 1999.

[39] ——, "Generalized network formulation analysis of the N–furcations application to elliptical waveguide," *Proc. 10th Int. Symp. on Theoretical Electrical Engineering, Magdeburg, Germany, (ISTET)*, pp. 129–134, Sep. 1999.

[40] M. Mongiardo, C. Tomassoni, and P. Russer, "Generalized network formulation: Application to flange–mounted radiating waveguides," *IEEE Transactions on Antennas and Propagation*, vol. 55, no. 6, pp. 1667–1678, jun 2007.

2

Representations of Electromagnetic Fields

I Introduction

The aim of this chapter is to *introduce the relevant equations* for the computation of electromagnetic fields and their network representations in a unified and systematic format. As is well known, *Maxwell's equations* provide the basic equations governing electromagnetic fields when complemented by *constitutive relations* pertaining to the media under consideration and by their relevant *boundary conditions*. These equations are suitable for initiating the numerical/analytical solution of the given problem.

When dealing with Maxwell's equations we shall emphasize the *Laplace ($\bar{s}$–domain) formulation* which has several advantages. First, $\bar{s}$–domain solutions are numerically efficient because once the solution is computed, frequency sweeps and transient analysis are also feasible with modest numerical effort. Second, $\bar{s}$–domain solutions are well suited to conversion into equivalent networks; these equivalent networks can be combined with external voltage and current sources, and the entire system can be modeled by using circuit simulators. Third, the electromagnetic analysis may be performed by using either differential- or integral-equation methods. In addition, there is the advantage of expressing the set of equations in a format that is common in the theory of linear systems. The format is such as to allow us to identify the state variables of the system, the sources, the observable quantities and all corresponding transfer functions. This approach also highlights issues concerning the uniqueness of the solution, the possibility of expressing the state of the system with a minimal amount of data, and the strategy for the applications of reduced–order models.

In this chapter we shall deal with *abstract representations* where the electromagnetic fields vary on a spatial and temporal continuum, i.e. with systems of infinite dimensions. This formalism can be adapted in later chapters to discretization and truncation processes in finite dimensions, making these systems suitable for numerical computations.

II Maxwell's Equations

Equations linking electromagnetic field quantities have been introduced by James Clerk Maxwell in an elegant treatise first published in 1873 and then inserted into [1] (see also [2] for more historical information). We shall assume that a student reader is familiar with these equations, which are usually introduced in preliminary courses, and that he/she has a general knowledge of the relevant experimental facts and their theoretical interpretation. In what follows, we summarize Maxwell's equations in the time, frequency and Laplace ($\bar{s}$) domains.

II.1 Maxwell's Equations in Time–Dependent Form

It is customary to write Maxwell's equations in either local or in global form; we shall first consider their local form. We also note that, unfortunately, it is customary to designate the local form as differential form and this generates some confusion with the general meaning that differential forms have. In the following of this book, since differential forms are not used, the ambiguity is resolved.

Local Form of Maxwell's Equations

In three–dimensional vector notation, with vector $\boldsymbol{r}$ indicating a position in space and t the time variable, Maxwell's equations are

$$\nabla \times \boldsymbol{E}(\boldsymbol{r},t) = -\frac{\partial \boldsymbol{B}(\boldsymbol{r},t)}{\partial t}\,, \qquad \text{Faraday's law} \tag{2.1a}$$

$$\nabla \times \boldsymbol{H}(\boldsymbol{r},t) = \frac{\partial \boldsymbol{D}(\boldsymbol{r},t)}{\partial t} + \boldsymbol{J}(\boldsymbol{r},t)\,, \qquad \text{Ampère's law} \tag{2.1b}$$

$$\nabla \cdot \boldsymbol{D}(\boldsymbol{r},t) = \rho_e(\boldsymbol{r},t)\,, \qquad \text{Gauss' law} \tag{2.1c}$$

$$\nabla \cdot \boldsymbol{B}(\boldsymbol{r},t) = 0\,, \qquad \text{Magnetic flux continuity} \tag{2.1d}$$

where bold face symbols denote vector quantities. The quantities are defined as

$\boldsymbol{E}(\boldsymbol{r},t)$	electric field strength
$\boldsymbol{D}(\boldsymbol{r},t)$	electric displacement
$\boldsymbol{B}(\boldsymbol{r},t)$	magnetic flux density
$\boldsymbol{H}(\boldsymbol{r},t)$	magnetic field strength
$\boldsymbol{J}(\boldsymbol{r},t)$	electric current density
$\rho_e(\boldsymbol{r},t)$	electric charge density

Equations (2.1a)–(2.1d) are not independent since, for example, we may derive (2.1d) by taking the divergence of (2.1a). Another fundamental relationship can be derived by introducing (2.1c) into the divergence of (2.1b)

$$\nabla \cdot \boldsymbol{J}(\boldsymbol{r},t) = -\frac{\partial \rho_e(\boldsymbol{r},t)}{\partial t} \tag{2.2}$$

which provides the conservation law for electric charge and current densities. Actually, the set of three equations (2.1a), (2.1b) and (2.2) may be considered as the independent equations describing macroscopic electromagnetic fields, since the two Gauss equations (2.1c) and (2.1d) can be derived from this set.
Note that in the static case $\frac{\partial}{\partial t} = 0$ the electric and magnetic fields are not any more interdependent and the equations (2.1a) – (2.1d) become

$$\nabla \times \boldsymbol{E}(\boldsymbol{r}) = 0\,, \tag{2.3a}$$

$$\nabla \times \boldsymbol{H}(\boldsymbol{r}) = \boldsymbol{J}(\boldsymbol{r})\,, \tag{2.3b}$$

$$\nabla \cdot \boldsymbol{D}(\boldsymbol{r}) = \rho_e(\boldsymbol{r})\,, \tag{2.3c}$$

$$\nabla \cdot \boldsymbol{B}(\boldsymbol{r}) = 0\,. \tag{2.3d}$$

Finally also note that, if we assign the electric current density $\boldsymbol{J}(\boldsymbol{r})$ and the electric charge density $\rho_e(\boldsymbol{r})$, we have, from (2.1a) and (2.1b), two vector equations (i.e. six scalar equations) while we have four unknown vectors (i.e. twelve scalar quantities). To complete the number of equations we have to account for the media properties expressed by the constitutive relations.

Integral (global) Form of Maxwell's Equations

The properties of an electromagnetic field may also be expressed globally by an equivalent system of integral relations through use of the two fundamental theorems of vector analysis: the divergence theorem and Stokes' theorem [3].

Divergence or Gauss' Theorem

Let $\boldsymbol{U}(\boldsymbol{r})$ be any vector function of position, continuous together with its first derivative throughout a volume V bounded by a surface S. The divergence theorem states that

$$\oint_S \boldsymbol{U}(\boldsymbol{r}) \cdot \boldsymbol{n}\, dS = \int_V \nabla \cdot \boldsymbol{U}(\boldsymbol{r})\, dV\,, \tag{2.4}$$

where $\mathbf{n}$ is the outward unit vector normal to S. In fact, Gauss's theorem may also be used to *define* the divergence.

Stokes' Theorem

Let $\boldsymbol{U}(\boldsymbol{r})$ be any vector function of position, continuous together with its first derivatives throughout an arbitrary surface S bounded by a contour C, and assumed to be resolvable into a finite number of regular arcs. Stokes' theorem (also called curl theorem) states that

$$\oint_C \boldsymbol{U}(\boldsymbol{r}) \cdot d\ell = \int_S [\nabla \times \boldsymbol{U}(\boldsymbol{r})] \cdot \boldsymbol{n}\, dS\,, \tag{2.5}$$

where $d\ell$ is an element of length along C, and $\boldsymbol{n}$ is a unit vector normal to the positive side of the element area dS as defined by the right–hand thumb rule. This relationship may also be considered as an equation defining the *curl* or *circulation.*

By applying the curl theorem to (2.1a) and (2.1b), and the divergence theorem to (2.1c) and (2.1d), we get

$$\oint_C \boldsymbol{E}(\boldsymbol{r},t) \cdot d\ell = -\int_S \frac{\partial \boldsymbol{B}(\boldsymbol{r},t)}{\partial t} \cdot \boldsymbol{n}\, dS\,, \tag{2.6a}$$

$$\oint_C \boldsymbol{H}(\boldsymbol{r},t) \cdot d\ell = \int_S \frac{\partial \boldsymbol{D}(\boldsymbol{r},t)}{\partial t} \cdot \boldsymbol{n}\, dS + \int_S \boldsymbol{J}(\boldsymbol{r},t) \cdot \boldsymbol{n}\, dS\,, \tag{2.6b}$$

$$\int_V \nabla \cdot \boldsymbol{D}(\boldsymbol{r},t) dV = \oint_S \boldsymbol{D}(\boldsymbol{r},t) \cdot \boldsymbol{n}\, dS = \int_V \rho_e(\boldsymbol{r},t)\, dV\,, \tag{2.6c}$$

$$\int_V \nabla \cdot \boldsymbol{B}(\boldsymbol{r},t) dV = \oint_S \boldsymbol{B}(\boldsymbol{r},t) \cdot \boldsymbol{n}\, dS = 0\,. \tag{2.6d}$$

By defining the current $I(t)$ as

$$I(t) = \int_S \boldsymbol{J}(\boldsymbol{r},t) \cdot \boldsymbol{n}\, dS, \tag{2.7}$$

the charge $Q(t)$ as

$$Q(t) = \int_V \rho_e\, dV, \tag{2.8}$$

and the flux of the magnetic induction as

$$\Phi_m(t) = \int_S \boldsymbol{B}(\boldsymbol{r},t) \cdot \boldsymbol{n}\, dS, \tag{2.9}$$

we may write the previous equations as

$$\oint_C \boldsymbol{E}(\boldsymbol{r},t) \cdot d\ell = -\frac{\partial \Phi_m(t)}{\partial t}\,, \tag{2.10a}$$

$$\oint_C \boldsymbol{H}(\boldsymbol{r},t) \cdot d\ell = \int_S \frac{\partial \boldsymbol{D}(\boldsymbol{r},t)}{\partial t} \cdot \boldsymbol{n}\, dS + I(t)\,, \tag{2.10b}$$

$$\int_V \nabla \cdot \boldsymbol{D}(\boldsymbol{r},t)\, dV = Q(t)\,. \tag{2.10c}$$

II.2 Maxwell's Equations in the Frequency Domain

Electromagnetic fields operating at a particular frequency are known as time–harmonic steady–state or monochromatic fields. By adopting the time dependence $e^{j\omega t}$ to denote a time–harmonic field with angular frequency ω, we write

$$\boldsymbol{E}(\boldsymbol{r},t) = \Re\left\{\boldsymbol{E}(\boldsymbol{r})e^{j\omega t}\right\}, \tag{2.11}$$

where $\Re$ denotes the mathematical operator which selects the real part of a complex quantity. The complex quantity is $\boldsymbol{E}(\boldsymbol{r})$ is called a *vector phasor*. In (2.11) we have used the same symbol to denote both the real quantity in the time domain, $\boldsymbol{E}(\boldsymbol{r},t)$, and the complex quantity, $\boldsymbol{E}(\boldsymbol{r})$, in the frequency domain. In what follows we shall generally refer to complex quantities unless otherwise explicitly stated.

By applying (2.11) to the field quantities appearing in (2.1a), (2.1b), (2.1c) and (2.1d) we obtain Maxwell's equations in the frequency domain. As an example, let us consider (2.1a) for which we have

$$\Re\left\{\left[\nabla \times \boldsymbol{E}(\boldsymbol{r}) + j\omega\boldsymbol{B}(\boldsymbol{r})\right]e^{j\omega t}\right\} = 0\,. \tag{2.12}$$

Since this equation is valid for *all times t*, we may make use of the above lemma and state that the quantity inside the square bracket must be equal to zero. By applying the same reasoning also to the other equations (2.1b), (2.1c) and (2.1d) we get

$$\nabla \times \boldsymbol{E}(\boldsymbol{r}) = -j\omega\boldsymbol{B}(\boldsymbol{r})\,, \tag{2.13a}$$
$$\nabla \times \boldsymbol{H}(\boldsymbol{r}) = j\omega\boldsymbol{D}(\boldsymbol{r}) + \boldsymbol{J}(\boldsymbol{r})\,, \tag{2.13b}$$
$$\nabla \cdot \boldsymbol{D}(\boldsymbol{r}) = \rho_e(\boldsymbol{r})\,, \tag{2.13c}$$
$$\nabla \cdot \boldsymbol{B}(\boldsymbol{r}) = 0\,. \tag{2.13d}$$

In the following, we make use of equivalence theorems which introduce magnetic current density, $\boldsymbol{M}(\boldsymbol{r})$, and magnetic charge distributions, $\rho_m(\boldsymbol{r})$. These quantities, although not physically present, help in the solution of several boundary value problems. When considering also magnetic currents and charges, the frequency–domain Maxwell's equations become

$$\nabla \times \boldsymbol{E}(\boldsymbol{r}) = -j\omega\boldsymbol{B}(\boldsymbol{r}) - \boldsymbol{M}(\boldsymbol{r})\,, \tag{2.14a}$$
$$\nabla \times \boldsymbol{H}(\boldsymbol{r}) = j\omega\boldsymbol{D}(\boldsymbol{r}) + \boldsymbol{J}(\boldsymbol{r})\,, \tag{2.14b}$$
$$\nabla \cdot \boldsymbol{D}(\boldsymbol{r}) = \rho_e(\boldsymbol{r})\,, \tag{2.14c}$$
$$\nabla \cdot \boldsymbol{B}(\boldsymbol{r}) = -\rho_m(\boldsymbol{r})\,. \tag{2.14d}$$

II.3 Maxwell's Equations in the $\bar{s}$–Domain

By introducing the complex variable $\bar{s} = \sigma + j\omega$, the Laplace transform is defined conventionally as

$$\boldsymbol{E}(\boldsymbol{r}, \bar{s}) = \int_0^\infty \boldsymbol{E}(\boldsymbol{r}, t) e^{-\bar{s}t} dt \,. \tag{2.15}$$

In (2.15) we have used the same symbol to denote both the quantity in the time domain, $\boldsymbol{E}(\boldsymbol{r}, t)$, and that in the $\bar{s}$-domain, $\boldsymbol{E}(\boldsymbol{r}, \bar{s})$. In what follows we shall generally refer to these quantities without explicitly exhibiting the $\bar{s}$ or t dependence, the latter being clear from the context.
Applying (2.15) to the field quantities appearing in (2.1a), (2.1b), (2.1c) and (2.1d) yields Maxwell's equations in the $\bar{s}$–domain,

$$\nabla \times \boldsymbol{E}(\boldsymbol{r}) = -\bar{s}\boldsymbol{B}(\boldsymbol{r}) - \boldsymbol{M}(\boldsymbol{r}) \,, \tag{2.16a}$$

$$\nabla \times \boldsymbol{H}(\boldsymbol{r}) = \bar{s}\boldsymbol{D}(\boldsymbol{r}) + \boldsymbol{J}(\boldsymbol{r}) \,, \tag{2.16b}$$

$$\nabla \cdot \boldsymbol{D}(\boldsymbol{r}) = \rho_e(\boldsymbol{r}) \,, \tag{2.16c}$$

$$\nabla \cdot \boldsymbol{B}(\boldsymbol{r}) = -\rho_m(\boldsymbol{r}) \,. \tag{2.16d}$$

II.4 Constitutive Relations

As already pointed out Maxwell's equations cannot be solved unless the relationships between the field vectors $\boldsymbol{D}$ and $\boldsymbol{B}$ with $\boldsymbol{E}$ and $\boldsymbol{H}$ are specified. The type of field generated by given sources depends on the medium characteristics, which are accounted for by *constitutive relations*; they may be written as

$$\boldsymbol{D} = \mathcal{F}_d(\boldsymbol{E}, \boldsymbol{H}) \,, \tag{2.17a}$$

$$\boldsymbol{B} = \mathcal{F}_b(\boldsymbol{E}, \boldsymbol{H}) \,. \tag{2.17b}$$

Here, $\mathcal{F}_d$ and $\mathcal{F}_d$ are suitable functionals dependent on the medium considered; they may be classified as:

- *nonlinear*, when functionals depend on the electromagnetic field;
- *inhomogeneous*, when functionals depend on *space coordinates*; they are called *spatially–dispersive* when functionals also depend on *spatial derivatives*;
- *nonstationary*, if functionals depend on *time* or *temporally–dispersive* when functionals depend on *time derivatives.*

We shall deal only with linear, stationary media; however, inhomogeneous media are included because of their practical importance.
Another classification of media is provided by the vector form of the constitutive relations. The simplest possibility arises when considering *isotropic media,* where the constitutive relations are given by

$$\boldsymbol{D} = \varepsilon \boldsymbol{E}, \tag{2.18a}$$
$$\boldsymbol{B} = \mu \boldsymbol{H}. \tag{2.18b}$$

with ε denoting permittivity and μ permeability. In this case $\boldsymbol{E}$ is parallel to $\boldsymbol{D}$ and $\boldsymbol{B}$ is parallel to $\boldsymbol{H}$. In particular, in free space, the above equations are rewritten by using the vacuum constitutive parameters, i.e. permittivity ε_0 and permeability μ_0, as

$$\boldsymbol{D} = \varepsilon_0 \boldsymbol{E}, \tag{2.19a}$$
$$\boldsymbol{B} = \mu_0 \boldsymbol{H}, \tag{2.19b}$$

with

$$\varepsilon_0 = 8.854 \cdot 10^{-12}\,\mathrm{Fm}^{-1} \cong \frac{1}{36\pi} 10^{-9}\,\mathrm{Fm}^{-1}, \tag{2.20a}$$
$$\mu_0 = 4\pi \cdot 10^{-7}\,\mathrm{Hm}^{-1}. \tag{2.20b}$$

Anisotropic media are characterized by constitutive relations of the type

$$\boldsymbol{D} = \underline{\underline{\varepsilon}}\, \boldsymbol{E}, \tag{2.21a}$$
$$\boldsymbol{B} = \underline{\underline{\mu}}\, \boldsymbol{H}, \tag{2.21b}$$

where $\underline{\underline{\mu}}$ is the permeability tensor and $\underline{\underline{\varepsilon}}$ is the permittivity tensor. The medium is called *electrically anisotropic* if it is described by the permittivity tensor $\underline{\underline{\varepsilon}}$, and *magnetically anisotropic* when it is described by the permeability tensor $\underline{\underline{\mu}}$. A medium can be both electrically and magnetically anisotropic. An interesting particular case is that of *biaxial* crystals, which may be described, by choosing a suitable particular coordinate system, the so–called principal system, in terms of a tensor of the type:

$$\underline{\underline{\varepsilon}} = \begin{bmatrix} \varepsilon_x & 0 & 0 \\ 0 & \varepsilon_y & 0 \\ 0 & 0 & \varepsilon_z \end{bmatrix}. \tag{2.22}$$

Cubic crystals, where $\varepsilon_x = \varepsilon_y = \varepsilon_z$, are isotropic; tetragonal, hexagonal and rhombohedral crystals have two parameters equal and the medium is called *uniaxial.* The *principal axis* that exhibits this anisotropy is also referred to as the *optic axis.* When all the three parameters are different, as in orthorhombic crystals, the medium is referred to as *biaxial.*

When the medium has elements possessing permanent electric and magnetic dipoles parallel or antiparallel to each other, an applied electric field simultaneously aligns electric *and* magnetic dipoles; analogously, an applied magnetic field that aligns the magnetic dipoles simultaneously aligns the electric dipoles [4, p.8]. In order to describe such media Tellegen, in 1948, introduced a new element, the *gyrator*, in addition to the resistor, the capacitor, the inductor and, the ideal transformer. These media, when placed in an electric field or a magnetic field, become both polarized and magnetized, and they are referred to as *bianisotropic*, being characterized by constitutive relations of the type

$$\boldsymbol{D} = \underline{\underline{\varepsilon}}\,\boldsymbol{E} + \underline{\underline{\xi}}\,\boldsymbol{H}\,, \tag{2.23a}$$

$$\boldsymbol{B} = \underline{\underline{\xi}}\,\boldsymbol{E} + \underline{\underline{\mu}}\,\boldsymbol{H}\,. \tag{2.23b}$$

Examples of hypothetic materials which directly relate electric and magnetic fields are the perfect electromagnetic conductors (PEMCs) as discussed by Sihvola and Lindell [5]. In a PEMC electric and magnetic fields on a material response level both cause electric and magnetic polarizations, however the medium response is not sensitive to the vector orientation of the electric and magnetic fields. P. Russer has introduced the field theoretical analogon to the gyrator circuit of network theory by boundary surfaces with gyrator properties [6].

II.5 Boundary Conditions

In order to obtain a unique solution of the Maxwell field equations, one must impose appropriate boundary, radiation, and edge conditions. Radiation and edge conditions formalize, respectively, the outgoing wave requirement on fields in an infinite region excited by sources in a bounded domain and by conservation of energy in the possibly singular fields induced in the vicinity of edges and corners (tips) on obstacle scatterers. These conditions are discussed customarily using field solutions of the wave equation in an appropriate coordinate system, and they are treated later on in Section VII. We shall deal here only with the boundary conditions arising at the interface between two different media.

Consider a regular surface S of a medium discontinuity, as shown in Figure 2.1, where the subscripts 1 and 2 distinguish quantities in regions 1 and 2, respectively. From (2.6a) and (2.6b), as a consequence of a limiting process, one obtains the following conditions:

$$\boldsymbol{n} \times (\boldsymbol{H}_2 - \boldsymbol{H}_1) = \boldsymbol{J}\,, \tag{2.24a}$$

$$\boldsymbol{n} \times (\boldsymbol{E}_2 - \boldsymbol{E}_1) = -\boldsymbol{M}\,, \tag{2.24b}$$

where $\boldsymbol{J}$ and $\boldsymbol{M}$ are, respectively, the electric and magnetic surface current density distributions at the interface. Similarly, from (2.6c) and (2.6d), for a small volume at the interface, a limiting process yields,

$$\boldsymbol{n} \cdot (\boldsymbol{B}_2 - \boldsymbol{B}_1) = -\rho_m\,, \tag{2.25a}$$

$$\boldsymbol{n} \cdot (\boldsymbol{D}_2 - \boldsymbol{D}_1) = \rho_e\,, \tag{2.25b}$$

where ρ_e and ρ_m are, respectively, the electric and magnetic surface charge density distributions on the interface.

If neither medium is perfectly conducting, the tangential component of the fields $\boldsymbol{E}$ and $\boldsymbol{H}$ are continuous while their normal components undergo a jump due to the discontinuity in the permittivity and permeability.
When medium 1 is a perfect electric conductor, the field inside the medium vanishes everywhere and induced electric charges and currents exist on the surface. In this case we have:

$$\boldsymbol{n} \times \boldsymbol{H}_2 = \boldsymbol{J}\,, \tag{2.26a}$$

$$\boldsymbol{n} \times \boldsymbol{E}_2 = 0\,, \tag{2.26b}$$

$$\boldsymbol{n} \cdot \boldsymbol{B}_2 = 0\,, \tag{2.26c}$$

$$\boldsymbol{n} \cdot \boldsymbol{D}_2 = \rho_e\,, \tag{2.26d}$$

which states the vanishing, at the metal surface, of the tangential components of $\boldsymbol{E}$ and of the normal component of $\boldsymbol{H}$.
In certain cases, it is convenient to include fields generated from equivalent magnetic currents. Accordingly, the field generated by a magnetic current distribution in the immediate vicinity of a perfectly (electrically) conducting surface is given by

$$\boldsymbol{n} \times \boldsymbol{E}_2 = -\boldsymbol{M}\,. \tag{2.27}$$

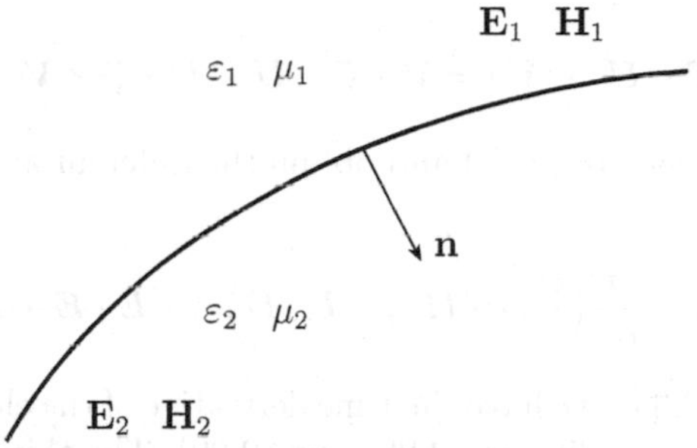

Fig. 2.1. Interface between two media.

III Theorems and Concepts for Electromagnetic Field Computation

Certain theorems and concepts of electromagnetic theory are of fundamental importance for efficient and systematic electromagnetic field computation. Their short description follows.

III.1 Energy and Power

The field concept is based upon the hypothesis that the electromagnetic energy is distributed over the space. We introduce the electric energy density

$$w_e(\boldsymbol{r},t) = \frac{\varepsilon}{2}\boldsymbol{E}(\boldsymbol{r},t)\cdot\boldsymbol{E}(\boldsymbol{r},t) \tag{2.28}$$

and the magnetic energy density

$$w_m(\boldsymbol{r},t) = \frac{\mu}{2}\boldsymbol{H}(\boldsymbol{r},t)\cdot\boldsymbol{H}(\boldsymbol{r},t). \tag{2.29}$$

In order to investigate energy storage and power flow in the electromagnetic field, we start again with Maxwell's equations. By scalar multiplication of Ampére's law with $-\boldsymbol{E}$ and Faraday's law with $\boldsymbol{H}$, we obtain

$$\begin{aligned} \nabla\times\boldsymbol{H} &= \tfrac{\partial\boldsymbol{D}}{\partial t}+\boldsymbol{J} \quad | \quad\cdot(-\boldsymbol{E}),\\ \nabla\times\boldsymbol{E} &= \quad -\tfrac{\partial\boldsymbol{B}}{\partial t} \quad | \quad\cdot\boldsymbol{H}. \end{aligned} \tag{2.30}$$

After inserting (2.18a) and (2.18b) into equation (2.30), we obtain

$$\boldsymbol{H}\cdot\nabla\times\boldsymbol{E}-\boldsymbol{E}\cdot\nabla\times\boldsymbol{H} = -\mu\boldsymbol{H}\cdot\frac{\partial\boldsymbol{H}}{\partial t}-\varepsilon\boldsymbol{E}\cdot\frac{\partial\boldsymbol{E}}{\partial t}-\boldsymbol{E}\cdot\boldsymbol{J}. \tag{2.31}$$

Using the relation

$$\nabla\cdot(\boldsymbol{U}\times\boldsymbol{V}) = \boldsymbol{V}\cdot\nabla\times\boldsymbol{U}-\boldsymbol{U}\cdot\nabla\times\boldsymbol{V}, \tag{2.32}$$

we transform the left side of (2.31) and obtain the differential form of *Poynting's theorem*

$$-\nabla\cdot(\boldsymbol{E}\times\boldsymbol{H}) = \frac{\partial}{\partial t}\left(\frac{\mu}{2}\boldsymbol{H}\cdot\boldsymbol{H}+\frac{\varepsilon}{2}\boldsymbol{E}\cdot\boldsymbol{E}\right)+\sigma\boldsymbol{E}\cdot\boldsymbol{E}+\boldsymbol{E}\cdot\boldsymbol{J}_0. \tag{2.33}$$

On the right side of (2.33), we have the time derivative of the electric and magnetic energy densities corresponding to (2.28) and (2.29). The third term is the power loss density

$$p_v(\boldsymbol{r},t) = \sigma(\boldsymbol{r})\boldsymbol{E}\cdot\boldsymbol{E}. \tag{2.34}$$

Due to the impressed current density $\boldsymbol{J}_0$, a power

$$p_0(\boldsymbol{r},t) = -\boldsymbol{E}(\boldsymbol{r},t)\cdot\boldsymbol{J}_0(\boldsymbol{r},t) \tag{2.35}$$

is added to the electromagnetic field per unit of volume. Introducing the *Poynting* vector

$$\boldsymbol{S}(\boldsymbol{r},t) = \boldsymbol{E}(\boldsymbol{r},t)\times\boldsymbol{H}(\boldsymbol{r},t) \tag{2.36}$$

allows to write down Poynting's theorem in the following form:

$$\nabla \cdot \boldsymbol{S} = -\frac{\partial w_m}{\partial t} - \frac{\partial w_e}{\partial t} - p_v + p_0 . \tag{2.37}$$

Integrating (2.37) over a volume V and transforming the integral over S into a surface integral over the boundary ∂V, we obtain the integral form of Poynting's theorem:

$$\oint_{\partial V} \boldsymbol{S} \cdot \mathrm{d}\boldsymbol{A} = \int_V p_0 \mathrm{d}V - \frac{\mathrm{d}}{\mathrm{d}t} \int_V w_m \mathrm{d}V - \frac{\mathrm{d}}{\mathrm{d}t} \int_V w_e \mathrm{d}V - \int_V p_v \mathrm{d}V . \tag{2.38}$$

The first term on the right side of equation (2.38) describes the power added into the volume V via impressed currents. The second and the third term, respectively, describe time variation of the magnetic and electric energy stored in the volume. The last term describes the conductive losses occurring inside the volume V. The right side of the equation comprises the total electromagnetic power generated within the volume V minus the power losses in the volume minus the increase of electric and magnetic power stored in the volume. This net power must be equal to the power, which is flowing out from the volume V through the boundary ∂V. Therefore we may interpret the surface integral over the pointing vector on the left side of (2.38) as the total power flowing from inside the volume V to the outside. Since this is valid for an arbitrary choice of volume V, it follows that the Poynting vector describes the energy flowing by unit of time through an unit area oriented perpendicular to S.

For harmonic electromagnetic fields, the introduction of a complex Poynting vector is useful. For this we construct

$$\begin{aligned} \nabla \times \boldsymbol{H}^* &= -j\omega\varepsilon^* \boldsymbol{E}^* + \boldsymbol{J}_0^* \quad &&| \ \cdot (-\boldsymbol{E}) , \\ \nabla \times \boldsymbol{E} &= \quad -j\omega\mu \boldsymbol{H} &&| \ \cdot \boldsymbol{H}^* . \end{aligned} \tag{2.39}$$

Summing both equations, we obtain

$$\boldsymbol{H}^* \cdot \nabla \times \boldsymbol{E} - \boldsymbol{E} \cdot \nabla \times \boldsymbol{H}^* = -j\omega(\mu \mid \boldsymbol{H} \mid)^2 - \varepsilon^* \mid \boldsymbol{E} \mid^2) - \boldsymbol{E} \cdot \boldsymbol{J}_0^* . \tag{2.40}$$

With the relation

$$\nabla \cdot (\boldsymbol{U} \times \boldsymbol{V}) = \boldsymbol{V} \cdot \nabla \times \boldsymbol{U} - \boldsymbol{U} \cdot \nabla \times \boldsymbol{V} , \tag{2.41}$$

we can transform (2.40) into the differential form of the *complex Poynting's theorem*

$$\boldsymbol{\nabla} \cdot \frac{1}{2}(\boldsymbol{E} \times \boldsymbol{H}^*) = -2j\omega \left(\frac{\mu}{4} \mid \boldsymbol{H} \mid^2 - \frac{\varepsilon^*}{4} \mid \boldsymbol{E} \mid^2 \right) - \frac{1}{2} \boldsymbol{E} \cdot \boldsymbol{J}_0^* . \tag{2.42}$$

We now introduce the *complex Poynting vector* $\boldsymbol{T}$:

$$\boldsymbol{T}(\boldsymbol{r}) = \frac{1}{2} (\boldsymbol{E}(\boldsymbol{r}) \times \boldsymbol{H}^*(\boldsymbol{r})) . \tag{2.43}$$

We have to note that $\boldsymbol{T}$ is not the phasor corresponding to $\boldsymbol{S}$. Therefore we have used a different character to distinguish between the complex Poynting vector and the real Poynting vector. In order to give an interpretation of the complex Poynting vector $\boldsymbol{T}$, we compute first the time-dependent Poynting vector $\boldsymbol{S}$ for a harmonic electromagnetic field

$$\boldsymbol{E}(\boldsymbol{r},t) = \Re\left\{\boldsymbol{E}(\boldsymbol{r})\mathrm{e}^{\mathrm{j}\omega t}\right\} = \frac{1}{2}\left(\boldsymbol{E}(\boldsymbol{r})\mathrm{e}^{\mathrm{j}\omega t} + \boldsymbol{E}^*(\boldsymbol{r})\mathrm{e}^{-\mathrm{j}\omega t}\right), \tag{2.44a}$$

$$\boldsymbol{H}(\boldsymbol{r},t) = \Re\left\{\boldsymbol{H}(\boldsymbol{r})\mathrm{e}^{\mathrm{j}\omega t}\right\} = \frac{1}{2}\left(\boldsymbol{H}(\boldsymbol{r})\mathrm{e}^{\mathrm{j}\omega t} + \boldsymbol{H}^*(\boldsymbol{r})\mathrm{e}^{-\mathrm{j}\omega t}\right) \tag{2.44b}$$

we obtain

$$\boldsymbol{S}(\boldsymbol{r},t) = \frac{1}{2}\Re\left\{\boldsymbol{E}(\boldsymbol{r}) \times \boldsymbol{H}^*(\boldsymbol{r})\right\} + \frac{1}{2}\Re\left\{\boldsymbol{E}(\boldsymbol{r}) \times \boldsymbol{H}(\boldsymbol{r})\mathrm{e}^{2\mathrm{j}\omega t}\right\}. \tag{2.45}$$

The first term on the right side of (2.45) is equal to the real part of the complex Poynting vector $\boldsymbol{T}$ after equation (2.43). This term is independent of time. The second on the right-hand side of (2.45) oscillates with twice the frequency of the alternating electromagnetic field. The time average of this part vanishes. Therefore the real part of the complex Poynting vector $\boldsymbol{T}$ is the time average of the Poynting vector $\boldsymbol{S}$.

$$\overline{\boldsymbol{S}(\boldsymbol{r},t)} = \Re\{\boldsymbol{T}(\boldsymbol{r})\}. \tag{2.46}$$

The real part of the complex Poynting vector $\boldsymbol{T}$ denotes the power flowing through an unit area oriented perpendicular to $\boldsymbol{T}$. We write the time averages of the electric and magnetic energy densities $\overline{w}_e$ and $\overline{w}_m$ as

$$\overline{w}_e = \frac{\varepsilon}{2}\overline{\boldsymbol{E}(\boldsymbol{r},t) \cdot \boldsymbol{E}(\boldsymbol{r},t)} = \frac{\varepsilon'}{4} \mid \boldsymbol{E}(\boldsymbol{r}) \mid^2, \tag{2.47}$$

$$\overline{w}_m = \frac{\mu}{2}\overline{\boldsymbol{H}(\boldsymbol{r},t) \cdot \boldsymbol{H}(\boldsymbol{r},t)} = \frac{\mu'}{4} \mid \boldsymbol{H}(\boldsymbol{r}) \mid^2. \tag{2.48}$$

We have to consider that the quantities ε' and μ' in the complex representation correspond to the quantities ε and μ in the time-dependent formulation. From equations (2.29), (2.43) and (2.34) we obtain the average electric power dissipation density

$$\overline{p}_{ve} = \frac{1}{2}\sigma \mid \boldsymbol{E}(\boldsymbol{r}) \mid^2 = \frac{1}{2}\omega\varepsilon'' \mid \boldsymbol{E}(\boldsymbol{r}) \mid^2. \tag{2.49}$$

The introduction of the complex permittivity μ allows also to consider the magnetic losses. The average power dissipation density is given by

$$\overline{p}_v = \frac{w}{2}(\varepsilon'' \mid \boldsymbol{E}(\boldsymbol{r}) \mid^2 + \mu'' \mid \boldsymbol{H}(\boldsymbol{r}) \mid^2. \tag{2.50}$$

The complex power, which is added to the field due to the impressed current density $\boldsymbol{J}_0$ is given by

$$p_{s0} = -\frac{1}{2}\boldsymbol{E} \cdot \boldsymbol{J}_0^*. \tag{2.51}$$

The real part of p_{s0} equals the time average $\overline{p}_{s0}$ according to equation (2.37).

$$\overline{p}_0 = \Re\{p_{s0}\}. \tag{2.52}$$

The proof is similar to the one of (2.46). After inserting of (2.43), (2.47), (2.48), (2.50) and (2.51) into (2.42), we can write down the complex Poynting's theorem in the following form

$$\boldsymbol{\nabla} \cdot \boldsymbol{T} = -2j\omega(\overline{w}_m - \overline{w}_e) - \overline{p}_v + p_{s0}\,. \tag{2.53}$$

By integration over a volume V, we obtain the integral form of the complex Poynting's theorem

$$\oint_{\partial V} \boldsymbol{T} \cdot \mathrm{d}\boldsymbol{A} = \int_V p_{s0}\mathrm{d}V - 2j\omega \int_V (\overline{w}_m - \overline{w}_e)\mathrm{d}V - \int_V \overline{p}_v \mathrm{d}V\,. \tag{2.54}$$

We consider first the real part of (2.54).

$$\Re\left\{\oint_{\partial V} \boldsymbol{T} \cdot \mathrm{d}\boldsymbol{A}\right\} = \Re\left\{\int_V p_{s0}\mathrm{d}V\right\} - \int_V \overline{p}_v \mathrm{d}V\,. \tag{2.55}$$

The left side of (2.55) equals the active power radiated from inside the volume V through the boundary ∂V. On the right side of this equation, the first term denotes the power added via the impressed current density $\boldsymbol{J}_0$; the second term describes the conductive losses, the dielectric losses and the magnetic losses inside the volume V.
The imaginary part of (2.54) is

$$\Im\left\{\oint_{\partial V} \boldsymbol{T} \cdot \mathrm{d}\boldsymbol{A}\right\} = \Im\left\{\int_V p_{s0}\mathrm{d}V\right\} - 2\omega \int_V (\overline{w}_m - \overline{w}_e)\mathrm{d}V\,. \tag{2.56}$$

The first term on the right side gives the reactive power inserted into the volume V via the impressed current density $\boldsymbol{J}_0$. Let us first consider the case where the second term on the right side is vanishing. In this case we see that the left side of (2.56) denotes the power radiated from volume V. Since the volume V can be chosen arbitrarily, it follows that the imaginary part of the complex Poynting vector $\boldsymbol{T}$ describes the reactive power radiated through an unit area normally oriented to the vector $\boldsymbol{T}$.

The second term on the right side of (2.56) contains the product of the double angular frequency with the difference of the average stored magnetic and electric energies. This term yields no contribution, if the magnetic energy stored in the volume V equals the average electric energy stored in V. The magnetic energy as well as electric energy oscillates with an angular frequency 2ω. The energy is permanently converted between electric energy and magnetic energy. If the averages $\overline{w}_e$ and $\overline{w}_m$ are equal, electric and magnetic energies may be mutually converted completely. In this case the energy oscillates between electric and magnetic field inside the volume V. If the average electric and magnetic energies are not equal, energy as well oscillates between volume V and the space outside V. In this case there is a power flow between V and the outer region. For $\overline{w}_m > \overline{w}_e$ the reactive power flowing into volume V is positive, whereas for $\overline{w}_m < \overline{w}_e$ the reactive power flowing into V is negative.

III.2 Field Theoretic Formulation of Tellegen's Theorem

Tellegen's theorem states fundamental relations between voltages and currents in a network, and is of considerable versatility and generality in network theory [7–9]. A notable property of this theorem is that it is only based on Kirchhoff's current and voltage laws, i.e. on topological relationships, and that it is independent of the constitutive laws of the network. The same reasoning that leads from Kirchhoff's laws to Tellegen's theorem permits direct derivation of a field form of Tellegen's theorem from Maxwell's equations [9–11].
In order to derive Tellegen's theorem for partitioned electromagnetic structures, let us consider two cases based on the same partitioning but filled with different materials. The connection network is established via relating the tangential field components on both sides of the boundaries; since the connection network has zero volume, no field energy is stored therein. An important point for the following discussion is that the materials filling the subdomains may be completely different. Starting directly from Maxwell's equations we may derive for a closed volume V with boundary surface ∂V and normal unit vector $\boldsymbol{n}$ the following relation

$$\int_{\partial V} \boldsymbol{E}'(\boldsymbol{\rho},t') \times \boldsymbol{H}''(\boldsymbol{\rho},t'') \cdot \boldsymbol{n}\, d\mathcal{A} = -\int_V \mathbf{E}'(\mathbf{r},t') \cdot \mathbf{J}''(\mathbf{r},t'')dV$$
$$-\int_V \mathbf{E}'(\mathbf{r},t') \cdot \frac{\partial \mathbf{D}''(\mathbf{r},t'')}{\partial t''}\, dV - \int_V \mathbf{H}'(\mathbf{r},\mathbf{t}') \cdot \frac{\partial \mathbf{B}''(\mathbf{r},\mathbf{t}'')}{\partial t''}\, dV\,. \quad (2.57)$$

The single and double primes relate to the case of a different choice of sources, different material parameters and also a different time reference. For volumes V of zero measure, we obtain the following equation

$$\int_{\partial V} \boldsymbol{E}'(\boldsymbol{\rho},t') \times \boldsymbol{H}''(\boldsymbol{\rho},t'') \cdot \boldsymbol{n} d\mathcal{A} = 0\,. \quad (2.58)$$

The above equation may be considered as the field form of Tellegen's theorem. Since it applies to a volume of zero measure, it is independent of the domain equations.

III.3 Uniqueness Theorem

The uniqueness theorem indicates how a problem should be properly formulated in order to provide one and only one solution. Uniqueness of the solution is a consequence of the proper imposition of the boundary conditions, since overdetermination, i.e. too many boundary conditions, may lead to no solution for a given problem, while a lack of boundary conditions may lead to multiple solutions. For time–harmonic electromagnetic fields, the uniqueness theorem states that when the sources and the tangential components of the electric *or* magnetic field are specified over the *whole* boundary surface of a given region, then the solution within this region is unique. This is actually true only if the medium is slightly lossy; otherwise it is possible to have a multiplicity of solution as, for example, for a closed resonator.

The proof of the uniqueness theorem follows from considering two different solutions $\boldsymbol{E}_1$, $\boldsymbol{H}_1$ and $\boldsymbol{E}_2$, $\boldsymbol{H}_2$ in the volume V bounded by the surface S excited by the same system of sources. Let us define the difference fields $\delta\boldsymbol{E}$ and $\delta\boldsymbol{H}$ as

$$\delta\boldsymbol{E} = \boldsymbol{E}_1 - \boldsymbol{E}_2 \,, \tag{2.59a}$$

$$\delta\boldsymbol{H} = \boldsymbol{H}_1 - \boldsymbol{H}_2 \,. \tag{2.59b}$$

By linearity, and since the sources are the same, the difference fields satisfy the source–free Maxwell equations

$$\nabla \times \delta\boldsymbol{E} = -\, j\omega\mu\delta\boldsymbol{H} \,, \tag{2.60a}$$

$$\nabla \times \delta\boldsymbol{H} = j\omega\varepsilon\delta\boldsymbol{E} \,, \tag{2.60b}$$

where it has been assumed that the permittivity ε and the permeability μ are of the following form

$$\varepsilon = \varepsilon_r - j\varepsilon_i \,, \tag{2.61a}$$

$$\mu = \mu_r - j\mu_i \,, \tag{2.61b}$$

i.e. a small, positive, imaginary part is present. As noted in [4] the proof also holds when the imaginary parts are both negative. By scalar multiplication of (2.60a) by $\delta\boldsymbol{H}^*$ and of the complex conjugate of (2.60b) by $\delta\boldsymbol{E}$ we obtain

$$\nabla \cdot (\delta\boldsymbol{E} \times \delta\boldsymbol{H}^*) = j\omega\varepsilon^* |\delta\boldsymbol{E}|^2 - j\omega\mu|\delta\boldsymbol{H}|^2 \,. \tag{2.62}$$

The complex conjugate of (2.62) also holds, giving

$$\nabla \cdot (\delta \boldsymbol{E}^* \times \delta \boldsymbol{H}) = -j\omega\varepsilon |\delta \boldsymbol{E}|^2 + j\omega\mu^* |\delta \boldsymbol{H}|^2 . \tag{2.63}$$

By adding (2.62) and (2.63), integrating over the volume V and applying the divergence theorem, we recover

$$\oint_S (\delta \boldsymbol{E} \times \delta \boldsymbol{H}^* + \delta \boldsymbol{E}^* \times \delta \boldsymbol{H}) \cdot dS = -2\omega \int_V \left(\varepsilon_i |\delta \boldsymbol{E}|^2 + \mu_i |\delta \boldsymbol{H}|^2\right) dV . \tag{2.64}$$

When the tangential components of $\boldsymbol{E}$ or $\boldsymbol{H}$ coincide on the boundary surface S, i.e. when either $\delta \boldsymbol{E}$ or $\delta \boldsymbol{H}$ are zero on S, we have

$$\oint_S (\delta \boldsymbol{E} \times \delta \boldsymbol{H}^* + \delta \boldsymbol{E}^* \times \delta \boldsymbol{H}) \cdot dS = 0 . \tag{2.65}$$

In this case, the right–hand side of (2.64) is zero only if $\delta \boldsymbol{E}$ and $\delta \boldsymbol{H}$ are identically zero in the region V. This proves the theorem.
As a last remark, observe that for lossless structures, when we look for modal spectra, we are seeking resonant solutions. In this case, the uniqueness theorem does not apply and an infinity of solutions is present.

III.4 Equivalence Theorem

There are several forms in which to state the *equivalence theorem* [4, 12] and, in view of its importance in the solution of electromagnetic field problems, it seems appropriate to examine relevant issues in detail.
Let us consider a volume V bounded by a surface S separating the internal region, labeled as region 1, from the external region, labeled as region 2. Our objective in applying the equivalence theorem is to maintain the field in region 2 even when modifying the field in region 1. By so doing, we obtain a modified field problem which, at least in region 2, and only in region 2, is equivalent to the original one. We denote by $\boldsymbol{E}_1$, $\boldsymbol{H}_1$ and $\boldsymbol{E}_2$, $\boldsymbol{H}_2$ the original fields in regions 1 and 2, respectively, as shown in Figure 2.2. Now suppose that the field in region 1 is altered, thus changing the field in this region from $\boldsymbol{E}_1, \boldsymbol{H}_1$ to $\boldsymbol{E}_1', \boldsymbol{H}_1'$. In order to maintain the original field in region 2, we must insert equivalent magnetic and electric currents, $\boldsymbol{M}_s$ and $\boldsymbol{J}_s$, respectively, on the surface S such that

$$\boldsymbol{J}_s = \boldsymbol{n} \times (\boldsymbol{H}_2 - \boldsymbol{H}_1') , \tag{2.66a}$$

$$\boldsymbol{M}_s = -\, \boldsymbol{n} \times (\boldsymbol{E}_2 - \boldsymbol{E}_1') , \tag{2.66b}$$

as shown in Figure 2.3.

Love Equivalence Theorem

A particular case is to set the field in region 1 equal to zero. Thus we have the case shown in Figure 2.4 where the surface currents are now given by

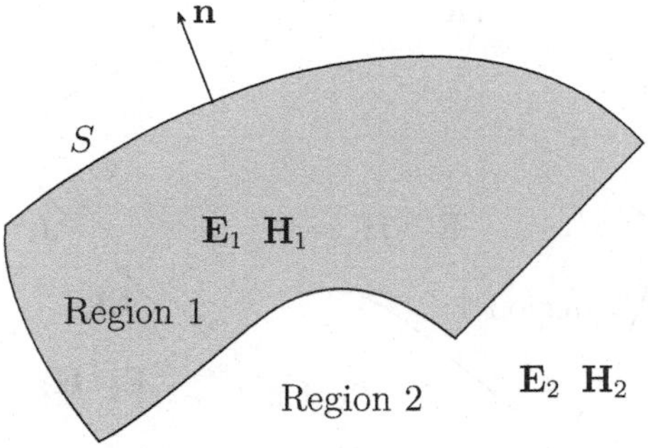

Fig. 2.2. Original field.

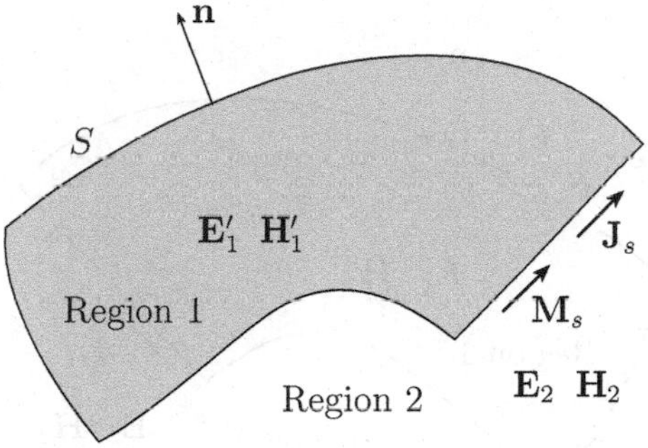

Fig. 2.3. The field in region 1 has been modified. By inserting the equivalent electric and magnetic currents on the surface S the field in region 2 is unchanged.

$$\boldsymbol{J}_s = \boldsymbol{n} \times \boldsymbol{H}_2\,, \tag{2.67a}$$

$$\boldsymbol{M}_s = -\,\boldsymbol{n} \times \boldsymbol{E}_2\,, \tag{2.67b}$$

The *Love equivalence theorem* states that the field in region 2 produced by the given sources in region 1 is the same as that produced by a system of *virtual* sources on the surface S.

Perfect Electric Conductor

The Love theorem only specifies a zero field in region 1. This may be obtained by filling region 1 with a perfect electric conductor (PEC) as considered here, or by filling region 1 with a perfect magnetic conductor (PMC) as considered below. It is easy to see that electric currents $\boldsymbol{J}_s$ on the PEC are short-circuited and therefore do not radiate any field. In fact, near the perfect conductor, the electric field is perpendicular to the surface S, while the magnetic field is parallel. The resulting Poynting vector $\boldsymbol{E} \times \boldsymbol{H}$ is thus parallel to the surface of the conductor and no

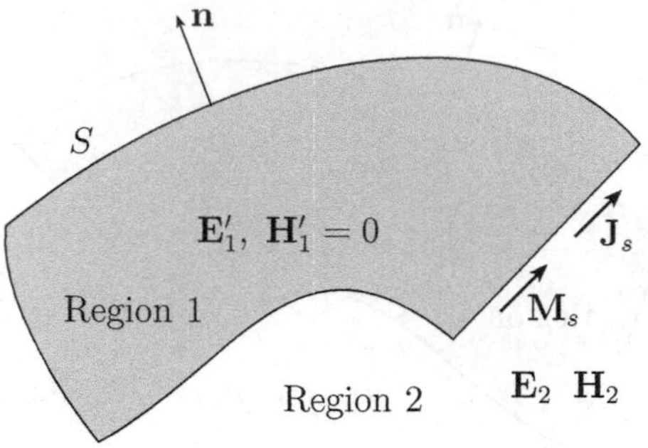

Fig. 2.4. The field in region 1 has been set to zero. Equivalent currents maintain the original field in region 2.

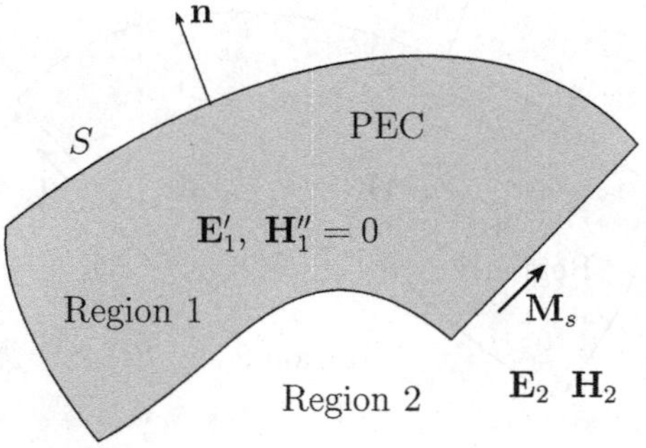

Fig. 2.5. Region 1 has been filled with a PEC. Only magnetic currents contribute since electric currents do not radiate.

power is radiated into space. A different proof of this fact may be obtained by using the Lorentz theorem. Thus, when region 1 is filled with a PEC, the resulting configuration is that shown in Figure 2.5. This form of equivalence theorem is used in practical applications where structures are bounded by metallic walls.

Perfect Magnetic Conductor

The other possibility of obtaining a null field in region 1, is to fill this region with a perfect magnetic conductor (PMC). In this case, since surface magnetic currents do not radiate, we are left with the case of Figure 2.6. Note that as in the previously, when calculating the field produced by sources in region 2, we must take into account the presence of the PMC, since the Green's function to be considered must satisfy the appropriate boundary conditions on the surface S. On the contrary, when applying the Love theorem without filling region 1

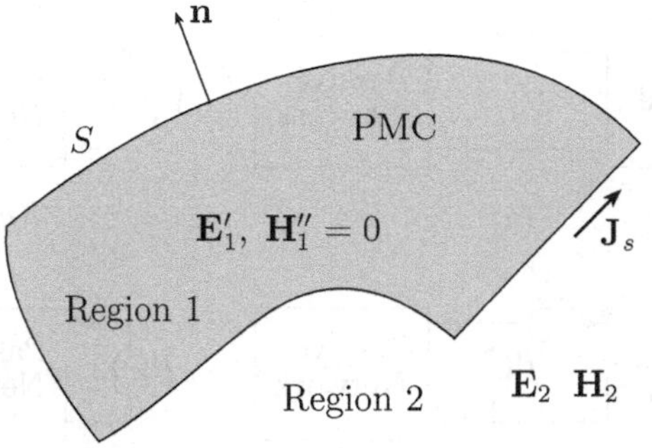

Fig. 2.6. Region 1 has been filled with a PMC Only electric currents contribute since mágnetic currents do not radiate.

with either a PMC or a PEC the Green's function to be considered is that of free–space.

The Circuit Theory Analog

The Circuit Theory analog of the equivalence theorem provides a simple and effective way to illustrate its utility [13]. Let region 2 be without sources, represented by the passive network in Figure 2.7(a), while region 1 is represented by the source–excited (active) network. We can set up an equivalent problem by

- switching off the sources in the active network, leaving the source impedance connected;
- placing a shunt current generator I equal to the terminal current in the original problem;
- placing a series voltage generator V equal to the terminal voltages into the original problem.

This replaces the original sources in region 1, the active network, by the virtual sources at the interface as shown in Figure 2.7(b). From conventional circuit concepts, it is evident that there is no excitation of the source impedance from these equivalent sources whereas the excitation of the passive network is unchanged. This fact offers the possibility of replacing the source impedance by either a short or an open circuit. By considering a short circuit, we obtain the case of Figure 2.7(c), equivalent to considering a PEC when applying the Love theorem. When using an open circuit, we obtain the case of Figure 2.7(d), equivalent to considering a PMC in the Love theorem.

Duality

Returning to the Maxwell equations in (2.14a)–(2.14d) it is noted that performing the substitutions

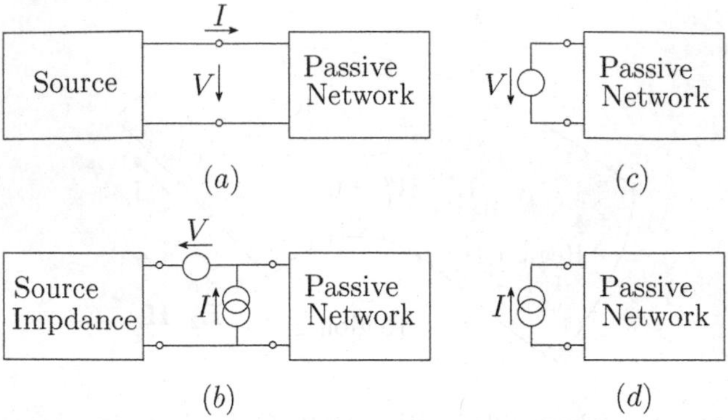

Fig. 2.7. Circuit analogue of the equivalence theorem: (a) original problem; (b) actual source deactivated, replaced by equivalent (virtual) sources; (c) source impedance replaced by a short circuit; (d) source impedance replaced by an open circuit.

$$\begin{aligned} &\boldsymbol{E} \to \boldsymbol{H} && \varepsilon \to \mu && \mu \to \varepsilon \\ &\boldsymbol{H} \to -\boldsymbol{E} && \boldsymbol{J} \to \boldsymbol{M} && \boldsymbol{M} \to -\boldsymbol{J}, \end{aligned} \tag{2.68}$$

equation (2.14a) becomes (2.14b), and vice versa. This is generally referred to as the "duality principle". However, the above substitutions imply a medium "dual" (or "adjoint") to free space, i.e. a medium with a permittivity of $4\pi \times 10^{-7}$ F/m and with permeability 8.854×10^{-12} H/m, which is undesirable.
A form of duality which is more suitable for antenna and radiation problems is established by the following equalities [4]

$$\begin{aligned} &\boldsymbol{E} \to \eta \boldsymbol{H}, && \boldsymbol{H} \to -\frac{1}{\eta}\boldsymbol{E}, \\ &\boldsymbol{J} \to \frac{1}{\eta}\boldsymbol{M}, && \boldsymbol{M} \to -\eta \boldsymbol{J}, \end{aligned} \tag{2.69}$$

with $\eta = \sqrt{\frac{\varepsilon}{\mu}}$ the free space impedance. With the substitutions in (2.69), equation (2.14a) becomes (2.14b) and vice versa, without need to replace free space with a different medium. Now, the form of duality in (2.69) does not apply to anisotropic or bianisotropic media, while (2.68) is more general.

IV Field Potentials

Auxiliary potentials are conventionally introduced to simplify the solution of the vector field equations [3, 14, 15]. When only *electric* sources are present in a homogeneous region, the two curl equations

$$\nabla \times \boldsymbol{E} = -j\omega\mu\boldsymbol{H}\,, \tag{2.70a}$$

$$\nabla \times \boldsymbol{H} = j\omega\varepsilon\boldsymbol{E} + \boldsymbol{J} \tag{2.70b}$$

provide six scalar equations. The divergence equations

$$\nabla \cdot \boldsymbol{D} = \rho_e\,, \tag{2.71a}$$

$$\nabla \cdot \boldsymbol{B} = 0\,, \tag{2.71b}$$

provide two additional scalar equations, which need to be complemented by the constitutive relations

$$\boldsymbol{D} = \varepsilon\boldsymbol{E}\,, \tag{2.72a}$$

$$\boldsymbol{B} = \mu\boldsymbol{H}\,, \tag{2.72b}$$

and the relevant boundary conditions. The use of potential functions can systematize the solution of this large set of equations.

Magnetic Vector and Electric Scalar Potentials

The *vector* and *scalar potential* functions $\boldsymbol{A}$ and Φ, respectively, represent the electrodynamic extensions of the static magnetic vector potential and electrical scalar potential, respectively. While potential theory is generally developed for the time–dependent form of Maxwell's equations [3, 14, 15], we shall deal directly with the time–harmonic potentials (an $\exp(j\omega t)$ time–dependence is assumed and suppressed). By taking the divergence of (2.70a) we see that

$$\nabla \cdot \boldsymbol{H} = 0, \tag{2.73}$$

i.e. the divergence equation for $\boldsymbol{H}$ is automatically satisfied. This suggests expressing $\boldsymbol{H}$ as

$$\boldsymbol{H} = \frac{1}{\mu}\nabla \times \boldsymbol{A}\,, \tag{2.74}$$

where $\boldsymbol{A}$ is referred to as the *magnetic vector potential.* By inserting (2.74) into (2.70a) we note that

$$\nabla \times (\boldsymbol{E} + j\omega\boldsymbol{A}) = 0 \tag{2.75}$$

and since

$$\nabla \times \nabla\Phi = 0 \tag{2.76}$$

the vector $\boldsymbol{E}$ can be expressed as

$$\boldsymbol{E} = -j\omega\boldsymbol{A} - \nabla\Phi \tag{2.77}$$

with Φ denoting the *electric scalar potential.* By substitution of (2.74), (2.77) into (2.70b) and recalling the vector identity

$$\nabla \times \nabla \times \boldsymbol{A} = \nabla\nabla \cdot \boldsymbol{A} - \nabla^2\boldsymbol{A} \tag{2.78}$$

we obtain

$$\nabla\nabla \cdot \boldsymbol{A} - \nabla^2\boldsymbol{A} = k^2\boldsymbol{A} - j\omega\mu\varepsilon\nabla\Phi + \mu\boldsymbol{J}\,. \tag{2.79}$$

Lorentz Potentials

Equation (2.79) can be phrased in a different manner by selecting the as yet unspecified divergence (lamellar part) of $\boldsymbol{A}$. One possible choice is to satisfy the *Lorentz* or *gauge condition*,

$$\nabla \cdot \boldsymbol{A} = -j\omega\mu\varepsilon\Phi \tag{2.80}$$

which reduces (2.79) to the *vector Helmholtz equation*

$$\nabla^2 \boldsymbol{A} + k^2 \boldsymbol{A} = -\mu \boldsymbol{J}. \tag{2.81}$$

Taking the divergence of (2.77) and using (2.80) it follows that the scalar potential Φ satisfies the scalar Helmholtz equation

$$\nabla^2 \Phi + k^2 \Phi = -\frac{\rho_e}{\varepsilon}. \tag{2.82}$$

The electric and magnetic fields, when using the Lorentz condition (2.80), are

$$\boldsymbol{E} = -j\omega \boldsymbol{A} + \frac{\nabla\nabla \cdot \boldsymbol{A}}{j\omega\mu\varepsilon}, \tag{2.83a}$$

$$\boldsymbol{H} = \frac{1}{\mu}\nabla \times \boldsymbol{A}. \tag{2.83b}$$

Electric Vector and Magnetic Scalar Potentials

When only *magnetic* sources are present, the vector $\boldsymbol{E}$ has zero divergence. By duality, introducing an *electric vector potential* $\boldsymbol{F}$ and a *scalar potential* Ψ, we obtain [13, p.129]

$$\boldsymbol{E} = -\frac{1}{\varepsilon}\nabla \times \boldsymbol{F}, \tag{2.84a}$$

$$\boldsymbol{H} = -j\omega \boldsymbol{F} + \frac{\nabla\nabla \cdot \boldsymbol{F}}{j\omega\varepsilon\mu} \tag{2.84b}$$

with

$$\nabla^2 \boldsymbol{F} + k^2 \boldsymbol{F} = -\varepsilon \boldsymbol{M}, \tag{2.85a}$$

$$\nabla^2 \Psi + k^2 \Psi = -\frac{\rho_m}{\mu}, \tag{2.85b}$$

where $\boldsymbol{M}$ denotes magnetic currents and ρ_m magnetic charges. When *electric and magnetic* currents are present simultaneously in a linear system, we make use of superposition to obtain:

$$\boldsymbol{E} = -j\omega \boldsymbol{A} + \frac{\nabla\nabla \cdot \boldsymbol{A}}{j\omega\mu\varepsilon} - \frac{1}{\varepsilon}\nabla \times \boldsymbol{F}, \tag{2.86a}$$

$$\boldsymbol{H} = \frac{1}{\mu}\nabla \times \boldsymbol{A} - j\omega \boldsymbol{F} + \frac{\nabla\nabla \cdot \boldsymbol{F}}{j\omega\varepsilon\mu}. \tag{2.86b}$$

Hertz Potentials

Hertz vector potentials for electric and magnetic time-harmonic fields are simply related to the electric and magnetic vector potentials via

$$\boldsymbol{A} = j\omega\varepsilon\mu\boldsymbol{\Pi}_e, \tag{2.87a}$$

$$\boldsymbol{F} = j\omega\varepsilon\mu\boldsymbol{\Pi}_h. \tag{2.87b}$$

The general field expression in terms of the Hertz vector potentials is:

$$\boldsymbol{E} = k^2\boldsymbol{\Pi}_e + \nabla\nabla\cdot\boldsymbol{\Pi}_e - j\omega\mu\nabla\times\boldsymbol{\Pi}_h, \tag{2.88a}$$

$$\boldsymbol{H} = j\omega\varepsilon\nabla\times\boldsymbol{\Pi}_e + k^2\boldsymbol{\Pi}_h + \nabla\nabla\cdot\boldsymbol{\Pi}_h. \tag{2.88b}$$

For Hertz potentials related to time-dependent fields see [15].

V Separation of Variables: The Scalar Wave Equation

Explicit solution of wave problems is facilitated substantially in special configurations that render the relevant wave equations fully or partially separable. Much about the physics of wave phenomena is learned from such special *canonical problems.* This section introduces concepts and notation for the *scalar wave equation*

$$\nabla^2\varphi - \frac{1}{c_0^2}\frac{\partial^2\varphi}{\partial t^2} = f(\mathbf{r},t), \tag{2.89}$$

where c_0 is the ambient wave speed. If it is assumed that the time dependence of the source distribution $f(\mathbf{r},t)$ is sinusoidal with frequency ω, the scalar field $\varphi(\mathbf{r},t)$ can be written as

$$\varphi(\mathbf{r},t) = \Re\left[U(u,v,w)e^{j\omega t}\right], \tag{2.90}$$

where (u,v,w) have been introduced as spatial coordinates. Outside the source region, (2.89) then becomes the *homogeneous Helmholtz equation*

$$(\nabla^2 + k_0^2)U = 0, \tag{2.91}$$

for the complex function U, where the ambient wavenumber k_0 is

$$k_0 = \frac{\omega}{c_0}, \tag{2.92}$$

with c_0 denoting the propagation speed. This equation is to be solved subject to the prevailing boundary conditions.

In certain spatial coordinate systems (see [16] for a complete discussion), it is possible to apply the *separation of variables* technique to (2.91), by which the field $U(u,v,w)$ is written as the product of functions which individually depend on only one spatial variable,

$$U(u,v,w) = U_u(u)U_v(v)U_w(w) \tag{2.93}$$

(see Table 2.1). Separability must hold for the operator ∇^2 *and* for the prevailing boundary conditions. Separability of the operator implies that the partial differential Laplacian ∇^2 can be arranged into three second-order one-dimensional ordinary differential operators ∇_τ^2, where τ stands for either u, v or w. On the boundaries $\tau = \tau_1$ and $\tau = \tau_2$ of each τ-domain, the boundary conditions are assumed to be of the linear homogeneous (impedance) type

$$U_\tau(\tau_{1,2}) + \gamma_{\tau_{1,2}} \left.\frac{\partial U_\tau(\tau)}{d\tau}\right|_{\tau_{1,2}} = 0 \quad \text{on} \quad B_{\tau_{1,2}} \tag{2.94}$$

Here, U_τ is one of the functions in (2.93) depending only on τ, B_τ defines the boundary surfaces $\tau =$ const. in the τ-domain, and $\gamma_{\tau 1,2}$ are constants. The method is best illustrated by example.

Table 2.1. Summary of boundary conditions for coordinate-separable solutions of the scalar wave equation.

Generic coordinates:	$\boldsymbol{r} = (u,v,w)$		
boundaries along:	$u = u_1, u_2,$	$v = v_1, v_2,$	$w = w_1, w_2,$
range of variables:	$u_1 \le u \le u_2,$	$v_1 \le v \le u_2,$	$w_1 \le w \le w_2$

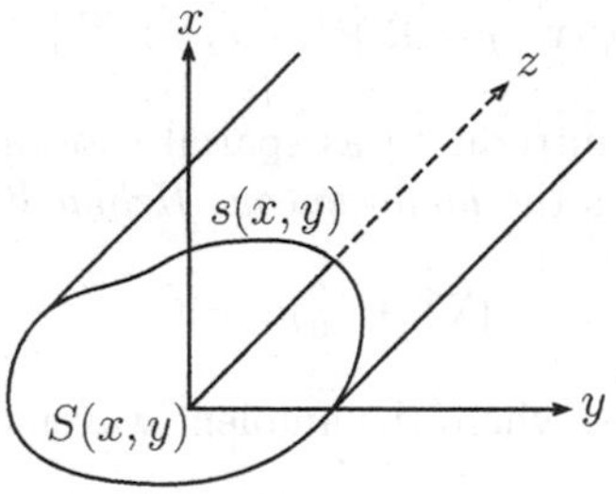

Fig. 2.8. Cross section $S(x,y)$ and boundary curve $s(x,y)$ for z-domain separation in Cartesian coordinates.

Table 2.2. Completely coordinate-separable configurations for rectangular Cartesian coordinates $(u, v, w) = (x, y, z)$. Note that, although only the cases for different conditions in the z–domain are shown, it is possible to change in a similar way also the x and y domains obtaining several other configurations.

Domain	Configuration
$x_1 = 0, x_2 = a,$ $y_1 = 0, y_2 = b,$ $z_1 = 0, z_2 = d,$	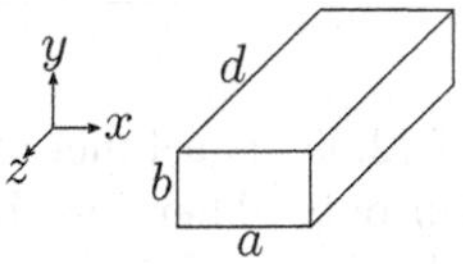
$x_1 = 0, \quad x_2 = a,$ $y_1 = 0, \quad y_2 = b,$ $z_1 \to -\infty, z_2 = 0,$	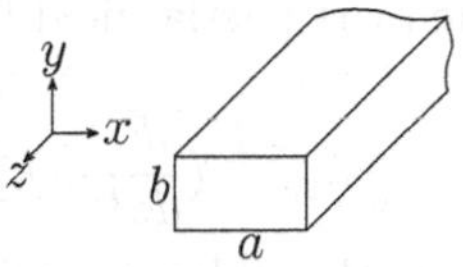
$x_1 = 0, \quad x_2 = a,$ $y_1 = 0, \quad y_2 = b,$ $z_1 \to -\infty, z_2 \to \infty,$	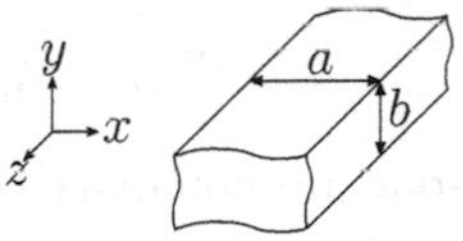

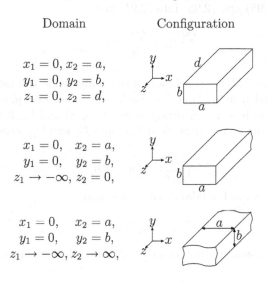

V.1 The Scalar Wave Equation in Cartesian Coordinates

The simplest demonstration of separability is for Cartesian coordinates, where $(u, v, w) = (x, y, z)$ (see Table 2.2). For a problem separable with respect to one of the coordinates, to be designated by z, the Laplacian ∇^2 is decomposed as

$$\nabla^2 = \nabla^2_{xy} + \nabla^2_z , \tag{2.95}$$

where

$$\nabla^2_{xy} = \frac{\partial^2}{\partial x^2} + \frac{\partial^2}{\partial y^2} \tag{2.96}$$

and

$$\nabla^2_z = \frac{\partial^2}{\partial z^2} . \tag{2.97}$$

Accordingly, a solution for the field $U(u, v, w)$ is sought in the form

$$U(x, y, z) = U_z(z) U_{xy}(x, y), \quad z_1 \leq z \leq z_2, \tag{2.98}$$

wherein the z-variable has been explicitly separated, and its domain has been explicitly identified. Any boundaries in the (x, y) domain must be z-independent cylindrical surfaces with transverse-to-z cross sections $S = S(x, y)$ bounded by

the curve $s(x, y)$ (Figure 2.8). Separability of the z-domain boundary conditions implies that the boundary conditions on the planes $z = z_1$ and $z = z_2$ must be independent of x and y, as in (2.94), with $\tau = z$. Similarly, the boundary conditions on the surface $s(x, y)$ must be independent of z.
Substitution of (2.95) and (2.98) into (2.91) gives

$$-\frac{\nabla_z^2 U_z}{U_z} = \frac{\nabla_{xy}^2 U_{xy}}{U_{xy}} + k_0^2. \tag{2.99}$$

The left-hand side of (2.99) is a function of z only, while the right-hand side is a function of x and y only. Therefore both sides must be equal to a constant, because (2.99) must hold for arbitrary values (x, y, z) and k_0. If this constant is designated by k_z^2, then the partial field functions U_z and U_{xy} satisfy the reduced equations

$$(\frac{d^2}{dz^2} + k_z^2)U_z = 0, \tag{2.100}$$

subject to the z-domain boundary conditions, and

$$(\nabla_{xy}^2 + k_{xy}^2)U_{xy} = 0, \tag{2.101}$$

subject to the (x, y)-domain boundary conditions. The constants k_z and k_{xy} satisfy the relation

$$k_0^2 - k_z^2 - k_{xy}^2 = 0. \tag{2.102}$$

With respect to wave propagation, k_0 is the total wavenumber, and the separation constants k_z and k_{xy} are therefore the wavenumber components associated with the z and the (x, y) subdomains, respectively. (2.102) is the *spatial dispersion relation* that constrains the wavenumber components.
When boundary conditions as in (2.94) are imposed at the endpoints z_1 and z_2 of the domain $z_1 \le z \le z_2$, solutions of (2.100) can be found for special values $k_z = k_{z\alpha}$, with corresponding solutions $U_{z\alpha}$. The problem

$$\left(\frac{d^2}{dz^2} + k_{z\alpha}^2\right) U_{z\alpha}(z) = 0\,, \qquad z_1 \le z \le z_2 \tag{2.103}$$

is called an *eigenvalue problem*, the wavenumbers (separation constants) $k_{z\alpha}$ are called *eigenvalues*, and the solutions $U_{z\alpha}(z)$ are called *eigenfunctions*. Depending on the boundary conditions at z_1 and z_2, the eigenvalues $k_{z\alpha}$ may form an infinite set of discrete values or they may be continuously distributed. Eigenvalue problems are discussed in detail within the context of the *Sturm–Liouville problem* (see Section VI).
If the boundary conditions are also (x, y)-separable (i.e., the three-dimensional problem is completely separable), the field U_{xy} in (2.98) is written as $U_{xy}(x, y) = U_x(x)U_y(y)$, so that the total scalar field $U(x, y, z) = U_{xy}U_z$ becomes

$$U(x,y,z) = U_x(x)U_y(y)U_z(z). \tag{2.104}$$

The individual partial fields U_x, U_y, U_z then satisfy the equations

$$(\frac{d^2}{dx^2} + k_x^2)U_x = 0\,, \tag{2.105a}$$

$$(\frac{d^2}{dy^2} + k_y^2)U_y = 0\,, \tag{2.105b}$$

$$(\frac{d^2}{dz^2} + k_z^2)U_z = 0,\,, \tag{2.105c}$$

with separation constants (wavenumbers) k_x, k_y and k_z that satisfy the dispersion relation

$$k_0^2 - k_x^2 - k_y^2 - k_z^2 = 0. \tag{2.106}$$

Imposition of the boundary conditions of (2.94) in the separate x and y domains leads to two eigenvalue problems analogous to that in (2.103), with corresponding interpretations. The solutions for $U_{z\alpha}$ are the trigonometric functions

$$U_{z\alpha}(z) = \sin k_{z\alpha}z, \quad \cos k_{z\alpha}z, \quad e^{ik_{z\alpha}z}, \quad e^{-ik_{z\alpha}z}, \tag{2.107}$$

any two of which are linearly independent, and are chosen in configurations that satisfy (2.94) at $z = z_{1,2}$. Solutions for $U_y(y)$ and $U_z(z)$ are similar.

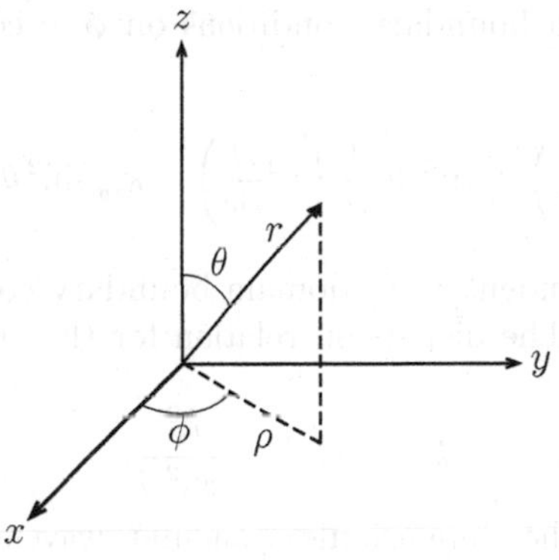

Fig. 2.9. Completely coordinate-separable configurations for spherical coordinates $(u, v, w) = (r, \theta, \phi)$. The domains $r = r_1, r = r_2$ correspond to spherical boundaries, $\theta = \theta_1, \theta = \theta_2$ to conical boundaries and $\phi = \phi_1, \phi = \phi_2$ to plane boundaries.

V.2 The Scalar Wave Equation in Spherical and Polar Coordinates

In separable curvilinear coordinates, the reduction process is similar but more subtle, because not all of the coordinates are direct measures of length. For example, in spherical polar coordinates $(u, v, w) = (r, \theta, \phi)$ (Figure 2.9), the Laplace operator is given by

$$\nabla^2 = \frac{1}{r^2}\frac{\partial}{\partial r}\left(r^2\frac{\partial}{\partial r}\right) + \frac{1}{r^2\sin^2\theta}\frac{\partial^2}{\partial\phi^2} + \frac{1}{r^2\sin\theta}\frac{\partial}{\partial\theta}\left(\sin\theta\frac{\partial}{\partial\theta}\right). \tag{2.108}$$

The azimuthal ϕ-coordinate is separated readily because the associated partial differential operator is $\partial^2/\partial\phi^2$ just as in the Cartesian case. Removing the coefficient $(r^2\sin^2\theta)^{-1}$ by cross multiplication, i.e., writing $U(r,\theta,\phi) = U_\phi(\phi)U_{r\theta}(r,\theta)$, (2.91) and (2.108) yields

$$-\frac{1}{U_\phi}\frac{\partial^2 U_\phi}{\partial\phi^2} = \frac{1}{U_{r\theta}}\left[\sin\theta\frac{\partial}{\partial\theta}\left(\sin\theta\frac{\partial}{\partial\theta}\right) + \sin^2\theta\frac{\partial}{\partial r}\left(r^2\frac{\partial}{\partial r}\right)\right]U_{r\theta} + k_0^2r^2\sin^2\theta. \tag{2.109}$$

The left-hand side of (2.109) is a function of ϕ only, while the right-hand side is a function of r and θ only. Therefore both sides must be equal to a constant, and if this constant is denoted by k_ϕ^2, then the functions U_ϕ and $U_{r\theta}$ satisfy the reduced equations

$$\left(\frac{d^2}{d\phi^2} + k_\phi^2\right)U_\phi = 0, \tag{2.110}$$

subject to the ϕ-domain boundary conditions on ϕ = const. planes, or for 2π-periodic conditions, and

$$\left(\sin\theta\frac{\partial}{\partial\theta}\left(\sin\theta\frac{\partial}{\partial\theta}\right) + \sin^2\theta\frac{\partial}{\partial r}\left(r^2\frac{\partial}{\partial r}\right) + k_{r\theta}^2\sin^2\theta\right)U_{r\theta} = 0, \tag{2.111}$$

subject to the ϕ-independent (r,θ)-domain boundary conditions on surfaces of revolution $S(r,\theta) = 0$. The dispersion relation for the angular wavenumbers k_ϕ and $k_{r\theta}$ is

$$k_{r\theta}^2 = k_0^2r^2 - \frac{k_\phi^2}{\sin^2\theta}. \tag{2.112}$$

Note that k_ϕ here is the dimensionless angular wavenumber associated with the dimensionless angular azimuthal coordinate ϕ. The separation parameter (wavenumber) $k_{r\theta}$ in (2.112) has been chosen so that the r and θ dependencies in $k_{r\theta}$ appear in separable form. If the problem conditions are also (r,θ) separable, then writing $U_{r\theta}(r,\theta) = U_r(r)U_\theta(\theta)$ in (2.111) gives

$$-\frac{1}{U_\theta}\frac{1}{\sin\theta}\frac{\partial}{\partial\theta}\left(\sin\theta\frac{\partial}{\partial\theta}\right)U_\theta + \frac{k_\phi^2}{\sin^2\theta} = \frac{1}{U_r}\frac{\partial}{\partial r}\left(r^2\frac{\partial}{\partial r}\right)U_r + r^2k_0^2. \tag{2.113}$$

Both sides of (2.113) must again be constant, and if this constant is denoted by k_θ^2, the partial fields U_θ and U_r satisfy the reduced equations

$$\left(\frac{1}{\sin\theta}\frac{d}{d\theta}\left(\sin\theta\frac{d}{d\theta}\right)-\frac{k_\phi^2}{\sin^2\theta}+k_\theta^2\right)U_\theta=0, \tag{2.114a}$$

$$\left(\frac{1}{r^2}\frac{d}{dr}\left(r^2\frac{d}{dr}\right)-\frac{k_\theta^2}{r^2}+k_0^2\right)U_r=0. \tag{2.114b}$$

Here, k_θ plays the role of the dimensionless wavenumber associated with the dimensionless angular latitudinal coordinate θ.
Solutions $U_\phi(\phi)$ of (2.110) are the trigonometric functions (see (2.107))

$$U_\phi(\phi)=\sin\mu\phi,\quad \cos\mu\phi,\quad e^{i\mu\phi},\quad e^{-i\mu\phi},\qquad \mu=k_\phi\,. \tag{2.115}$$

Solutions $U_\theta(\theta)$ of (2.114a) are the associated Legendre functions, with a linearly independent pair given by

$$U_\theta(\theta)=P_\nu^{-\mu}(\cos\theta),\quad P_\nu^{-\mu}(-\cos\theta),$$

$$\nu=\sqrt{k_\theta^2+\frac{1}{4}}-\frac{1}{2},$$

$$\text{i.e.,}\quad k_\theta^2=\nu(\nu+1) \tag{2.116}$$

and μ defined in (2.115). Solutions $U_r(r)$ of (2.114a) are the spherical Bessel functions

$$U_r(r)=j_\nu(k_0r)\,,\quad n_\nu(k_0r)\,,\quad h_\nu^{(1)}(k_0r)\,,\quad h_\nu^{(2)}(k_0r)\,, \tag{2.117}$$

with the order ν defined in (2.116). Any two of these solutions are linearly independent. The spherical Bessel functions are related to the cylindrical Bessel functions by

$$z_\nu(k_0r)=\sqrt{\frac{\pi}{2k_0r}}Z_{\nu+1/2}(k_0r)\,, \tag{2.118}$$

where z_ν stands for any of the spherical functions in (2.117) and $Z_{\nu+1/2}$ stands for the corresponding cylindrical Bessel function of argument (k_0r) and order $(\nu+1/2)$ (see (2.124)).

V.3 The Scalar Wave Equation in Cylindrical Polar Coordinates

In cylindrical polar coordinates $(u,v,w)=(\rho,\phi,z)$ (see Figure 2.10), the Laplacian operator is given by

$$\nabla^2=\frac{1}{\rho}\frac{\partial}{\partial\rho}\rho\frac{\partial}{\partial\rho}+\frac{1}{\rho^2}\frac{\partial^2}{\partial\phi^2}+\frac{\partial^2}{\partial z^2}\,. \tag{2.119}$$

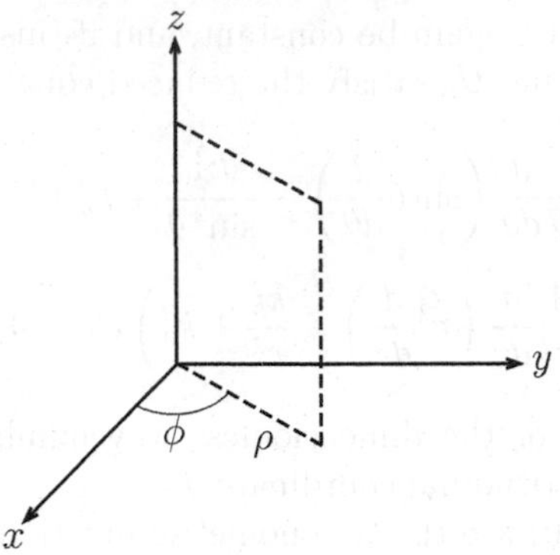

Fig. 2.10. Completely coordinate-separable configurations for cylindrical coordinates $(u, v, w) = (\rho, \phi, z)$. The domains $\rho = \rho_1, \rho = \rho_2$ correspond to cylindrical boundaries, $\phi = \phi_1, \phi = \phi_2$ and $z = z_1, z = z_2$ to plane boundaries.

Table 2.3. Some separable configurations in cylindrical coordinates. The figures show the ρ, ϕ plane.

$\rho_1 = 0$	$\rho_2 \to \infty$	$\rho_1 = 0$, $\rho_2 \to \infty$	radial section	angular periodic

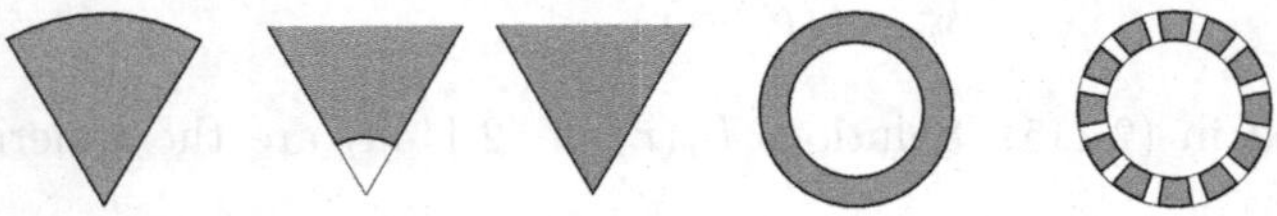

Writing $U(\rho, \phi, z) = U_z(z)U_{\rho\phi}(\rho, \phi)$ separates the z-dependence from the (ρ, ϕ)-dependence to yield, on substituting into (2.91) and proceeding as in (2.110) and (2.111), using k_z^2 as the separation parameter,

$$\left(\frac{d^2}{dz^2} + k_z^2\right) U_z(z) = 0\,, \tag{2.120}$$

$$\left(\frac{1}{\rho}\frac{\partial}{\partial \rho}\rho\frac{\partial}{\partial \rho} + \frac{1}{\rho^2}\frac{\partial^2}{\partial \phi^2} + k_0^2 - k_z^2\right) U_{\rho\phi}(\rho, \phi) = 0\,. \tag{2.121}$$

This decomposition applies to cylindrical boundaries of unchanged, but arbitrary, cross section along z, with non–separable boundary conditions (see Table 2.3). If the cross section is circular, the ρ- and ϕ-dependencies separate as well to yield, using k_ϕ^2 as the separation parameter,

$$\left(\frac{d^2}{d\phi^2} + k_\phi^2\right) U_\phi(\phi) = 0\,, \tag{2.122}$$

$$\left(\frac{1}{\rho}\frac{d}{d\rho}\rho\frac{d}{d\rho} + k_\rho^2\right) U_\rho(\rho) = 0\,, \tag{2.123}$$

where

$$k_\rho^2 = k_0^2 - k_z^2 - \frac{k_\phi^2}{\rho^2} \tag{2.124}$$

is the corresponding dispersion relation. Solutions $U_z(z)$ of (2.120) and $U_\phi(\phi)$ of (2.122) are trigonometric functions as in (2.107) and (2.115), respectively. Solutions $U_\rho(\rho)$ of (2.123) are cylindrical Bessel functions

$$U_\rho(\rho) = J_\mu(\kappa\rho)\,, \quad N_\mu(\kappa\rho)\,, \quad H_\mu^{(1)}(\kappa\rho)\,, \quad H_\mu^{(2)}(\kappa\rho)\,, \qquad \kappa = \sqrt{k_0^2 - k_z^2}\,, \tag{2.125}$$

any two of which are linearly independent. Here, $\mu = k_\phi$.

Table 2.3 summarizes the boundary configurations which allow a solution of the scalar wave equation by separation of variables. Figures 2.8, 2.9 and 2.10 show the corresponding domain configurations in Cartesian, spherical and cylindrical coordinate systems, respectively.

VI Sturm–Liouville Problems

VI.1 Source–Free Solutions: Eigenvalue Problem

Formulation

The reduced one-dimensional differential equations given by (2.105a)-(2.105c), (2.110), (2.114a), (2.114b), (2.120), (2.121), (2.122) and (2.123) all are special cases of the generic form

$$L_\alpha(u) f_\alpha(u) = 0, \tag{2.126}$$

$$L_\alpha(u) \equiv \left[\frac{d}{du}\left(p(u)\frac{d}{du}\right) - q(u) + \lambda_\alpha w(u)\right]\,, \tag{2.127}$$

where p, q and the weight function w are positive real functions of u, $f_\alpha(u)$ is the wave function, and λ_α is the separation parameter. Equation (2.127) is a homogeneous (source–free) *Sturm–Liouville* (SL) problem, [17, p.719] and $L(u)$ is the *Sturm–Liouville operator*, defined, in general, for arbitrary λ, as

$$L(u,\lambda) \equiv \left[\frac{d}{du}\left(p(u)\frac{d}{du}\right) - q(u) + \lambda w(u)\right]\,. \tag{2.128}$$

(2.127) is to be solved on the interval $u_1 \leq u \leq u_2$, subject to the linear homogeneous boundary conditions at the end points u_1 and u_2 (see (2.94)),

$$f_\alpha(u_{1,2}) + \gamma_{1,2}\left.\frac{df_\alpha}{du}\right|_{u_{1,2}} = 0, \tag{2.129}$$

where $\gamma_{1,2}$ are constants. As already noted in SectionV, a solution $f_\alpha(u)$ is called an *eigenfunction*, and the constant λ_α associated with $f_\alpha(u)$ is the corresponding *eigenvalue*. In general there will be a set of eigenfunction-eigenvalue pairs $\{(f_\alpha, \lambda_\alpha)\}$ which satisfy (2.127) and the boundary conditions in 2.129. Note that here and in the mathematical sections that follow, the *spectral* parameter λ in (2.128) plays a general role which, in the context of the wave equation, is equivalent to the squared wavenumber k^2. For the *eigenvalue problem* in (2.126), $\lambda \to \lambda_\alpha$ and $L(u, \lambda) \to L_\alpha(u)$.

Adjointness Properties

Before proceeding further, we demonstrate that the Sturm–Liouville (SL) operator $L(u)$ in (2.128) is *self-adjoint*; i.e., subject to the boundary conditions in (2.129), with f_α replaced by $F(u)$, $L(u)$ exhibits the adjointness property (suppressing the λ dependence)

$$\langle \bar{F}, LF \rangle \equiv \int_{u_1}^{u_2} \bar{F} LF \, du = \int_{u_1}^{u_2} FL\bar{F} \, du \equiv \langle F, L\bar{F} \rangle \,. \tag{2.130}$$

Equation (2.130) states that in the domain $u_1 \le u \le u_2$ of the operator $L(u)$ with the boundary conditions as in (2.129), the *inner product* $\langle \bar{F}, LF \rangle$ as defined on the left-hand side of (2.130) is equal to $\langle F, L\bar{F} \rangle$ on the right-hand side. The function $\bar{F}(a)$ is said to be adjoint to $F(a)$. Thus, the L-operation in the inner product is commutative. To prove (2.130), we construct

$$FL\bar{F} = F\left(\frac{d}{du}p\frac{d}{du} - q + \lambda w\right)\bar{F} \,, \tag{2.131}$$

$$\bar{F}LF = \bar{F}\left(\frac{d}{du}p\frac{d}{du} - q + \lambda w\right)F \,, \tag{2.132}$$

whence

$$\begin{aligned} FL\bar{F} - \bar{F}LF &= F\frac{d}{du}p\frac{d\bar{F}}{du} - \bar{F}\frac{d}{du}p\frac{dF}{du} \\ &= \frac{d}{du}\left[p\left(F\bar{F}' - \bar{F}F'\right)\right] \,. \end{aligned} \tag{2.133}$$

Here and hereafter, $F(u)$ and $\bar{F}(u)$ are two different twice–differentiable functions of u with a prime denoting the derivative with respect to u. Integrating both sides of (2.133) between the limits u_1 and u_2 yields

$$\int_{u_1}^{u_2} du\,\left(FL\bar{F} - \bar{F}LF\right) = \left[p\left(F\bar{F}' - \bar{F}F'\right)\right]_{u_1}^{u_2} \,. \tag{2.134}$$

The bracketed term on the right-hand side of (2.134),

$$W(\bar{F}, F) \equiv p\,(F\bar{F}' - \bar{F}F') = p \det \begin{vmatrix} F & F' \\ \bar{F} & \bar{F}' \end{vmatrix} \tag{2.135}$$

is the λ-dependent *Wronskian* which plays an important role in the theory that follows (see (2.186) - (2.196)). Subject to the boundary conditions in (2.129), the Wronskian vanishes, thereby establishing (2.130). Vanishing of the Wronskian at the boundary is confirmed by noting that, in view of (2.129),

$$F_1' = -\frac{F_1}{\gamma_1}\,, \qquad \bar{F}_1' = -\frac{\bar{F}_1}{\gamma_1}\,, \tag{2.136}$$

where $F_1 \equiv F(u_1)$, $\bar{F}_1 \equiv \bar{F}(u_1)$. The same holds for u_2.

Orthogonality, Completeness Relation, and Eigenfunction Expansions

In view of the adjointness property in (2.130), the eigenfunctions f_α satisfy an orthogonality property which can be derived as follows. Equation (2.127) is written for an eigenfunction-eigenvalue pair $(f_\alpha, \lambda_\alpha)$ and for a different eigenfunction-eigenvalue pair (f_β, λ_β). Proceeding as in (2.131) to (2.133), the f_α-equation is multiplied by f_β^*, where the asterisk denotes the complex conjugate, and the complex conjugate of the f_β-equation is multiplied by f_α. The resulting equations are subtracted to obtain

$$\frac{d}{du}W(f_\alpha, f_\beta^*) + (\lambda_\alpha - \lambda_\beta^*)w f_\alpha f_\beta^* = 0\,. \tag{2.137}$$

(2.137) is now integrated with respect to u between u_1 and u_2 to give

$$(\lambda_\alpha - \lambda_\beta^*) \int_{u_1}^{u_2} w f_\alpha f_\beta^*\, du = 0\,, \tag{2.138}$$

since the endpoint contribution vanishes via (2.136). Therefore it follows that

$$\int_{u_1}^{u_2} f_\alpha f_\beta^* w\, du = 0, \quad \alpha \neq \beta\,. \tag{2.139}$$

If $\lambda_\beta = \lambda_\alpha$, then from (2.138),

$$(\lambda_\alpha - \lambda_\alpha^*) \int_{u_1}^{u_2} |f_\alpha|^2 w\, du = 0\,. \tag{2.140}$$

Since w is positive and the trivial eigenfunction $f_\alpha = 0$ is not considered, the integral is nonzero. Thus,

$$\lambda_\alpha = \lambda_\alpha^*, \text{ i.e., } \textit{the eigenvalues are real}. \tag{2.141}$$

Returning to (2.139), these considerations imply that the integral vanishes for $\lambda_\alpha \neq \lambda_\beta$. Since the integral in (2.139) represents, in the function space, the inner product of the functions f_α and f^*_β, (see the comments following (2.130)), vanishing of the integral implies that eigenfunctions corresponding to distinct eigenvalues are *orthogonal* with respect to the weighting function w. It can also be shown that the eigenvalues λ_α are non-negative when γ_1 is negative real and γ_2 is positive real. In (2.131), let $\bar{F} = f^*_\alpha$, $F = f_\alpha$ and $\lambda = \lambda_\alpha$, and equate the expression to zero in view of (2.127). Integrating over the interval from u_1 to u_2 yields

$$\int_{u_1}^{u_2} du\, f_\alpha \frac{d}{du} p \frac{df^*_\alpha}{du} - \int_{u_1}^{u_2} du\, q\,|f_\alpha|^2 + \lambda_\alpha \int_{u_1}^{u_2} du\, w\,|f_\alpha|^2 = 0\,. \tag{2.142}$$

Integrating by parts, the first integral in (2.142) becomes

$$\int_{u_1}^{u_2} du\, f_\alpha \frac{d}{du}\left(p f^{*\prime}_\alpha\right) = p f_\alpha f^{*\prime}_\alpha\Big|_{u_1}^{u_2} - \int_{u_1}^{u_2} du\, p\,\left|f^{*\prime}_\alpha\right|^2. \tag{2.143}$$

For the boundary conditions in (2.129), with $\gamma_1 < 0, \gamma_2 > 0$ and real, the endpoint contributions at u_1 and u_2 can be written as $(-p|f_\alpha|^2|\gamma_1|^{-1})$ and $(-p|f_\alpha|^2\gamma_2^{-1})$, respectively. Thus, since $p > 0$, the left-hand side of (2.143), and therefore the first term in (2.142), is negative. The second term in (2.142) is also negative since $q > 0$, whereas the integral multiplying λ_α equals unity (see (2.144)). Thus, to satisfy (2.142), λ_α must be non-negative under the stated conditions.
It is convenient to normalize the eigenfunctions (multiply by an appropriate constant) so that

$$\int_{u_1}^{u_2} |f_\alpha|^2 w\, du = 1\,. \tag{2.144}$$

This renders the set $\{f_\alpha\}$ orthonormal. Equations (2.139) and (2.144) can then be written as the single expression

$$\int_{u_1}^{u_2} f_\alpha f^*_\beta w\, du = \delta_{\alpha\beta}\,, \tag{2.145}$$

with the Kronecker delta $\delta_{\alpha\beta}$ defined as $\delta_{\alpha\beta} = 0$ for $\alpha \neq \beta$ and $\delta_{\alpha\beta} = 1$ for $\alpha = \beta$. To evaluate the normalizing integral on the left side of (2.144) (i.e., *before* the normalization implied by the right side), we return to (2.137) but replace the eigenfunction $f^*_\beta(u) \equiv f^*_\beta(u, \lambda_\beta)$ by a function $f^*(u, \lambda)$ satisfying $L(u)f(u, \lambda) = 0$ (see (2.128)) for any specified $\lambda \neq \lambda_\alpha$. Integrating the resulting modification of (2.137) between the limits u_1 and u_2 yields

$$(\lambda_\alpha - \lambda) \int_{u_1}^{u_2} du\, w f_\alpha f^* + W(f_\alpha, f^*)\Big|_{u_1}^{u_2} = 0\,, \tag{2.146}$$

which can be re-arranged as follows,

$$\int_{u_1}^{u_2} du\, w f_\alpha f^* = \frac{W(f_\alpha, f^*)\Big|_{u_1}^{u_2}}{\lambda - \lambda_\alpha}\,. \tag{2.147}$$

Now take the limit $\lambda \to \lambda_\alpha$, whence $f^* \to f^*_\alpha$. The limiting form of the Wronskian vanishes, and the resulting indeterminate right-hand side can be evaluated by L'Hospital's rule, i.e., taking $[(dW/d\lambda)/(d/d\lambda)(\lambda - \lambda_\alpha)]_{\lambda=\lambda_\alpha}$, to obtain

$$\int_{u_1}^{u_2} du\, w(u)|f_\alpha(u)|^2 =$$

$$\left[p\left(\frac{d}{d\lambda}f^*(u,\lambda)\Big|_{\lambda_\alpha}\frac{d}{du}f_\alpha(u,\lambda_\alpha) - f_\alpha(u,\lambda_\alpha)\frac{d^2}{d\lambda du}\,f^*(u,\lambda)|_{\lambda_\alpha}\right)\right]_{u_1}^{u_2} \tag{2.148}$$

with

$$f(u,\lambda) \equiv f(\sqrt{\lambda}u)\,, \qquad f_\alpha(u,\lambda_\alpha) \equiv f(\sqrt{\lambda_\alpha}u)\,. \tag{2.149}$$

In (2.149), the functional dependencies of f and f_α are shown explicitly. The normalized eigenfunctions f_α defined in (2.144) are now obtained by writing $f_\alpha = B^{-1/2}\overline{f}_\alpha$, where $\overline{f}_\alpha$ is the unnormalized form, and $B = \{\ldots\}^{1/2}$, with $\{\ldots\}$ representing the expression on the right-hand side of (2.148) with (2.149).

Assuming that the eigenfunction set $\{f_\alpha(u)\}$ is complete, any "representable" function $F(u)$ can be expanded formally as

$$F(u) = \sum_\alpha A_\alpha f_\alpha(u)\,. \tag{2.150}$$

Here, "representable" implies that the expansion converges. Multiplying both sides of (2.150) by $w(u)f^*_\beta(u)$, integrating over the (u_1, u_2) interval, invoking the orthonormality condition given by (2.145) and switching back to the index α, it follows that

$$A_\alpha = \int_{u_1}^{u_2} f^*_\alpha F w\, du\,. \tag{2.151}$$

Substitution of (2.151) into (2.150) gives, upon interchange of the orders of summation and integration,

$$F(u) = \int_{u_1}^{u_2} du'\,\{w(u')\sum_\alpha f_\alpha(u)f^*_\alpha(u')\}F(u')\,, \tag{2.152}$$

which implies that

$$w(u')\sum_\alpha f_\alpha(u)f^*_\alpha(u') = \delta(u-u') \tag{2.153}$$

or

$$\frac{\delta(u-u')}{w(u')} = \sum_\alpha f_\alpha(u)f^*_\alpha(u')\,. \tag{2.154}$$

Equation (2.154) expresses the *completeness statement* in compact symbolic form. The expansion of the weighted delta function in terms of the eigenfunctions implies that the set of eigenfunctions is complete, because any function $F(u)$ can be expressed by using the delta function property

$$F(u) = \int_{u_1}^{u_2} F(u')\delta(u-u')\,du'\,. \tag{2.155}$$

Thus, to *apply* (2.154), the previous steps are *reversed* as follows. Each side of (2.154) is multiplied by $w(u')F(u')$ and integrated with respect to the variable u' from u_1 to u_2, giving

$$F(u) = \sum_\alpha f_\alpha(u) \int_{u_1}^{u_2} du'\, w(u')F(u')f_\alpha^*(u')\,. \tag{2.156}$$

Equation (2.156) is of the form

$$F(u) = \sum_\alpha A_\alpha f_\alpha(u)\,, \tag{2.157}$$

with the coefficients A_α given by

$$A_\alpha = \int_{u_1}^{u_2} du'\, w(u')F(u')f_\alpha^*(u')\,. \tag{2.158}$$

The implied orthonormality of the eigenfunctions is verified by setting $F(u)$ in (2.156) equal to the eigenfunction $f_\beta(u)$, giving

$$f_\beta(u) = \sum_\alpha f_\alpha(u) \int_{u_1}^{u_2} du'\, w(u')f_\beta(u')f_\alpha^*(u')\,. \tag{2.159}$$

To satisfy (2.159) one is led to (2.145).

Large $|\lambda|$ Behavior of the Source-Free Solutions

The source-free solutions $f(u)$ of the SL equation (see (2.127)) reduce to trigonometric functions for large values of λ, and when $w = p$. To demonstrate this behavior, we reduce the $L(u)$ operator to its normal form (without the first derivative d/du) by the transformation

$$f(u) = p^{-1/2}\hat{f}(u)\,, \tag{2.160}$$

which changes $L(u)f(u) = 0$ to the normalized equation

$$\left[\frac{d^2}{du^2} + h(u)\right]\hat{f}(u) = 0\,, \tag{2.161}$$

where

$$h(u) = \frac{\lambda w}{p} - \frac{q}{p} - p^{-1/2}\frac{d^2}{du^2}p^{1/2}\,. \tag{2.162}$$

For large λ, the $(\lambda w/p)$ term dominates, and when $w = p$, (2.161) reduces to

$$\left(\frac{d^2}{du^2}+\lambda\right)\hat{f}(u)\sim 0,\quad |\lambda|\gg 1,\quad w=p \tag{2.163}$$

Thus, the large-$|\lambda|$ solutions for $f(u)$ become

$$f(u)\sim p^{-1/2}\cdot\left(\sin\sqrt{\lambda}u\,,\quad \cos\sqrt{\lambda}u\,,\quad e^{\mp j\sqrt{\lambda}u}\right),\quad |\lambda|\gg 1\,,\quad w=p\,. \tag{2.164}$$

For the eigenvalue problem, (2.164) applies with $f(u)\to f_\alpha(u)$, $\lambda\to\lambda_\alpha$, $\lambda_\alpha\gg 1$. For the Green's function problem in Section VI.2, (2.164) applies to the synthesizing homogeneous solutions $f(\overleftarrow{u})$ and $f(\overrightarrow{u})$.

VI.2 Source-Driven Solutions: Green's Function Problem

Properties of the Green's Function

The eigenvalue problem defined by (2.126) describes a one-dimensional physical system which is free or unforced. Problems in which forcing functions or sources exist are solved through the introduction of a Green's function. The one-dimensional Green's function $g(u,u';\lambda)$ satisfies equation

$$L(u)g(u,u';\lambda)\equiv\left[\frac{d}{du}p(u)\frac{d}{du}-q(u)+\lambda w(u)\right]g(u,u';\lambda)=-\delta(u-u') \tag{2.165}$$

over the interval $u_1\le(u,u')\le u_2$, with boundary conditions at $u=u_{1,2}$ of the form (cf. VI.1)

$$g(u_{1,2})+\gamma_{1,2}\left.\frac{dg}{du}\right|_{u_{1,2}}=0\,. \tag{2.166}$$

The right-hand side of (2.165) represents a u-domain point source at location $u=u'$. Here, $L(u)$ is the general Sturm-Liouville (SL) operator in (2.128), which is self-adjoint subject to the boundary conditions in (2.129). The parameter λ is now unrestricted and may range over the entire complex λ-plane, provided that $\lambda\neq\lambda_\alpha$. All eigenvalues $\lambda=\lambda_\alpha$ must be avoided because the source-free (2.165) has the eigensolutions $f_\alpha(u)$. Any eigensolution can be added to g and still satisfy (2.165) and (2.166), thereby rendering the resulting g non unique.

Reciprocity

The Green's function $g(u,u';\lambda)$ is symmetric in its dependence on u and u'. This can be shown by referring to (2.134), with $F=g(u,u';\lambda)$ and $\bar{F}=g(u,u'';\lambda)$, where u' and u'' are source points in the interval $u_1<(u',u'')<u_2$. Thus (omitting the λ-dependence),

$$\int_{u_1}^{u_2}du\,[g(u,u')L(u)g(u,u'')-g(u,u'')L(u)g(u,u')]$$
$$=\{p(u)\,[g(u,u')g'(u,u'')-g(u,u'')g'(u,u')]\}_{u_1}^{u_2}\,. \tag{2.167}$$

Since $L(u)g(u,\overline{u}) = -\delta(u-\overline{u})$ and $g'(u_{1,2},\overline{u}) = -\gamma_{1,2}^{-1}g(u_{1,2},\overline{u})$ (see (2.165) and (2.166)), the endpoint contribution vanishes (self-adjointness property) and the integral is reduced via the delta functions, yielding the result

$$g(u'',u';\lambda) = g(u',u'';\lambda)\,. \tag{2.168}$$

Thus, the self-adjoint Sturm-Liouville (SL) Green's function $g(u,u';\lambda)$ is unchanged, i.e. *reciprocal* when u and u' are interchanged at any two locations in the interval (u_1,u_2).

Synthesis of the General Initial-Boundary Value Problem

The general SL initial-boundary value problem is of the form

$$L(u)F(u) = S(u)\,, \qquad u_1 \le u \le u_2\,, \tag{2.169}$$

subject to the initial-boundary condition

$$F(u_{1,2}) + \gamma_{1,2}F'(u_{1,2}) = \overline{S}(u_{1,2})\,, \tag{2.170}$$

where $S(u)$ are interior sources while $\overline{S}(u_{1,2})$ are sources impressed at the boundaries of the domain. The solution for $F(u)$ can be synthesized in terms of the Green's function $g(u,u';\lambda)$ defined in (2.165) together with (2.166). Returning to the adjointness relation in (2.133), let $\bar{F} = g(u,u';\lambda)$ and let $F(u)$ represent the solution of (2.169) and (2.170). Thus, omitting the λ-dependence,

$$\begin{aligned}\int_{u_1}^{u_2} du\,[F(u)L(u)g(u,u') - g(u,u')L(u)F(u)]&\\ = p(u_2)[F(u_2)g'(u_2,u') - g(u_2,u')F'(u_2)]&\\ - p(u_1)[F(u_1)g'(u_1,u') - g(u_1,u')F'(u_1)]\,.&\end{aligned} \tag{2.171}$$

Inside the integral in (2.171), referring to (2.165) and (2.169), Lg and LF are replaced by $-\delta(u-u')$ and $S(u)$, respectively. On the right-hand side of (2.171), referring to (2.166) and (2.170), we use

$$g'(u_{1,2}) = -\frac{g(u_{1,2})}{\gamma_{1,2}}\,, \qquad F'(u_{1,2}) = \frac{\overline{S}(u_{1,2}) - F(u_{1,2})}{\gamma_{1,2}}\,. \tag{2.172}$$

This reduces (2.171) to the expression

$$\begin{aligned}F(u') = -\int_{u_1}^{u_2} du\; g(u,u')S(u) + p(u_2)g(u_2,u')\overline{S}(u_2)/\gamma_2&\\ - p(u_1)g(u_1,u')\overline{S}(u_1)/\gamma_1&\end{aligned} \tag{2.173}$$

where u' is any point in the closed interval $u_1 \le u' \le u_2$.
Since u and u' in $g(u, u'; \lambda)$ represent the field (observation) point and source point, respectively, it is customary to integrate the Green's function over the primed coordinates. The necessary interchange of u and u' can be implemented in view of the reciprocity property in (2.168) in the form (restoring the λ-dependence)

$$F(u, \lambda) = -\int_{u_1}^{u_2} du'\, g(u, u'; \lambda) S(u') + p(u_2) g(u, u_2; \lambda) \overline{S}(u_2)/\gamma_2 \\ - p(u_1) g(u, u_1; \lambda) \overline{S}(u_1)/\gamma_1 \,. \tag{2.174}$$

Solution for the Green's Function

The Green's function $g(u, u'; \lambda)$ can be evaluated directly. When $u \neq u'$, the Green's function satisfies the homogeneous equation obtained by setting the right-hand side of (2.165) equal to zero. Let $\overleftarrow{f}$ be a solution of the homogeneous equation which satisfies the boundary condition given by (2.166) at $u = u_1$, and let $\overrightarrow{f}$ be a solution of the homogeneous equation which satisfies the boundary condition given by (2.166) at $u = u_2$. The functions $\overleftarrow{f}$ and $\overrightarrow{f}$ can be constructed by superposition of any two linearly independent solutions $f^{(1)}$ and $f^{(2)}$ of the homogeneous (2.126) using the expressions

$$\overrightarrow{f}(u) = f^{(1)}(u) + \overrightarrow{\Gamma} f^{(2)}(u) \,, \tag{2.175}$$

$$\overleftarrow{f}(u) = \overleftarrow{\Gamma} f^{(1)}(u) + f^{(2)}(u) \,, \tag{2.176}$$

where

$$\overrightarrow{\Gamma} = -\frac{\left[f^{(1)}(u_2) + \gamma_2 \left(\dfrac{df^{(1)}}{du} \right) \right]_{u=u_2}}{\left[f^{(2)}(u_2) + \gamma_2 \left(\dfrac{df^{(2)}}{du} \right) \right]_{u=u_2}} \,, \tag{2.177}$$

$$\overleftarrow{\Gamma} = -\frac{\left[f^{(2)}(u_1) + \gamma_1 \left(\dfrac{df^{(2)}}{du} \right) \right]_{u=u_1}}{\left[f^{(1)}(u_1) + \gamma_1 \left(\dfrac{df^{(1)}}{du} \right) \right]_{u=u_1}} \,. \tag{2.178}$$

To obtain the expression for $\overrightarrow{\Gamma}$, note from (2.166) that

$$\overrightarrow{f} = f^{(1)} + \overrightarrow{\Gamma} f^{(2)} = -\gamma_2 \frac{d\overrightarrow{f}}{du} = -\gamma_2 \left[\frac{df^{(1)}}{du} + \overrightarrow{\Gamma} \frac{df^{(2)}}{du} \right] , \qquad u = u_2 \,. \tag{2.179}$$

The second equality follows from (2.129) applied to $\vec{f}$, whereas the third equality implements $d\vec{f}/du$ via (2.175). Solving the first and third equalities for $\vec{\Gamma}$ yields (2.177). A similar calculation gives the expression for $\overleftarrow{\Gamma}$ in (2.178).
Next, it is noted that g is continuous at $u = u'$ but has a discontinuous slope (first derivative) at $u = u'$, consistent with the recognition that the delta function singularity at $u = u'$ in (2.165) is generated by the highest derivative, (d^2g/du^2). Implementing continuity at u', with discontinuous slope, suggests the expression

$$g(u,u';\lambda) = \begin{cases} \bar{C}\,\overleftarrow{f}(u)\,\vec{f}(u'), & u < u' \\ \bar{C}\,\overleftarrow{f}(u')\,\vec{f}(u), & u > u' \end{cases}, \tag{2.180}$$

which also satisfies both prescribed boundary conditions, as well as (2.165) for all $u \neq u'$. With the notation

$$u_> = \begin{cases} u, & u > u' \\ u', & u < u' \end{cases}, \tag{2.181}$$

$$u_< = \begin{cases} u, & u < u' \\ u', & u > u', \end{cases} \tag{2.182}$$

(2.180) can be written as

$$g(u,u';\lambda) = \bar{C}\,\vec{f}(u_>)\,\overleftarrow{f}(u_<)\,. \tag{2.183}$$

To determine the constant $\bar{C}$ we integrate (2.165) over the interval $u' - \epsilon < u < u' + \epsilon$, $\epsilon > 0$, and then allow $\epsilon \to 0$. Since g is bounded at u' and q, w and p have no singularities at $u = u'$, the contribution from the second and third terms in $L(u)$ vanishes in the limit. The result is

$$p\,\frac{dg}{du}\bigg|_{u'-\epsilon}^{u'+\epsilon} = -1\,, \tag{2.184}$$

which after using (2.183) gives

$$\bar{C} = -\frac{1}{W(\vec{f},\overleftarrow{f})}\,, \tag{2.185}$$

with the Wronskian $W(\vec{f},\overleftarrow{f})$ defined as in (2.135),

$$W(\vec{f},\overleftarrow{f}) = p(u')\left[\overleftarrow{f}\,\frac{d\vec{f}}{du} - \vec{f}\,\frac{d\overleftarrow{f}}{du}\right]_{u=u'}\,. \tag{2.186}$$

Using (2.183) and (2.185), the Green's function $g(u,u',\lambda)$ can now be written as

$$g(u,u';\lambda) = -\frac{\overrightarrow{f}\,(u_>)\,\overleftarrow{f}\,(u_<)}{W(\overrightarrow{f},\overleftarrow{f})}. \tag{2.187}$$

The Wronskian $W(\overrightarrow{f},\overleftarrow{f})$ has the following properties (recall that the λ–dependence has been suppressed throughout):

- W is a λ–dependent constant, independent of u'.
- $W \neq 0$ if $\overrightarrow{f}$ and $\overleftarrow{f}$ are linearly independent functions over the interval $u_1 < u < u_2$.

To show that W is independent of u', the equation

$$\left[\frac{d}{du}p\frac{d}{du} - q + \lambda w\right]\overrightarrow{f} = 0 \tag{2.188}$$

is multiplied by $\overleftarrow{f}$, and the equation

$$\left[\frac{d}{du}p\frac{d}{du} - q + \lambda w\right]\overleftarrow{f} = 0 \tag{2.189}$$

is multiplied by $\overrightarrow{f}$. The resulting equations are subtracted to give

$$\overleftarrow{f}\,\frac{d}{du}p\frac{d\overrightarrow{f}}{du} - \overrightarrow{f}\,\frac{d}{du}p\frac{d\overleftarrow{f}}{du} = 0, \tag{2.190}$$

which is equivalent to

$$\frac{d}{du}\left[p\left(\overleftarrow{f}\,\frac{d\overrightarrow{f}}{du} - \overrightarrow{f}\,\frac{d\overleftarrow{f}}{du}\right)\right] = 0, \tag{2.191}$$

or

$$\frac{d}{du}\left(W(\overrightarrow{f},\overleftarrow{f})\right) = 0. \tag{2.192}$$

Equation (2.192) states that W is independent of u, i.e., W equals a λ–dependent constant.

To show that W is nonzero if $\overrightarrow{f}$ and $\overleftarrow{f}$ are linearly independent, it will be shown conversely that $W = 0$ implies linear dependence, i.e., that $\overrightarrow{f}$ is then a constant multiple of $\overleftarrow{f}$. If $W = 0$, (2.186) gives

$$\overleftarrow{f}\,\frac{d\overrightarrow{f}}{du} = \overrightarrow{f}\,\frac{d\overleftarrow{f}}{du} \tag{2.193}$$

or

$$\frac{1}{\overleftarrow{f}}\frac{d\overleftarrow{f}}{du} = \frac{1}{\overrightarrow{f}}\frac{d\overleftarrow{f}}{du}. \tag{2.194}$$

Integration of (2.194) gives

$$\ln\frac{\overleftarrow{f}}{\overrightarrow{f}} = \bar{c} = \text{const.} \tag{2.195}$$

or

$$\overleftarrow{f} = \bar{c}'\,\overrightarrow{f} \tag{2.196}$$

which confirms that $W = 0$ implies linear dependence of $\overleftarrow{f}$ and $\overrightarrow{f}$. (2.196) implies furthermore that $\overrightarrow{f}$ or $\overleftarrow{f}$ satisfy *both* boundary conditions at u_1 and u_2, in addition to satisfying the source-free (2.165); i.e., $\overrightarrow{f}$ or $\overleftarrow{f}$ are eigensolutions $f_\alpha(u)$ with forbidden eigenvalues $\lambda = \lambda_\alpha$. This is in accord with the result in (2.136). Evidently, the solution for g in (2.187) becomes invalid when $W = 0$.

Large $|\lambda|$ Behavior of the Spectral Green's Function

In the investigation that follows, emphasis will be placed on the behavior of the Sturm–Liouville Green's function *throughout* the complex spectral $|\lambda|$–plane. For large values of λ, and when $w = p$, the synthesizing homogeneous solutions $\overrightarrow{f}$ and $\overleftarrow{f}$ in (2.187) reduce to trigonometric functions, as shown in Section VI.1, (2.164). The formal solution for $g(u, u'; \lambda)$ in (2.187) reduces accordingly in the large-λ range. Consider the case where $g = 0$ at $u_1 = 0$ (no loss of generality) and at u_2. The synthesizing solutions of (2.163) are $\overleftarrow{f}\ (u_<) = \sin\left(\sqrt{\lambda}u_<\right)$, $\overrightarrow{f}\ (u_>) = \sin\left[\sqrt{\lambda}(u_2 - u_>)\right]$, whereas the u–independent Wronskian is given by $w = \sqrt{\lambda}\sin(\sqrt{\lambda}u_2)$. For $\mid \lambda \mid\gg 1$, $\mid \Im\lambda \mid\neq 0$, retaining only the dominant (growing) exponentials, one obtains

$$g(u, u'; \lambda) \to \frac{e^{|\Im\sqrt{\lambda}|u_<}e^{|\Im\sqrt{\lambda}|(u_2-u_>)}}{\sqrt{\lambda}e^{|\Im\sqrt{\lambda}|u_2}} \to \frac{e^{|\Im\sqrt{\lambda}|(u_<-u_>)}}{\sqrt{\lambda}}$$

$$\frac{e^{-|\Im\sqrt{\lambda}|\,|u-u'|}}{\sqrt{\lambda}}, \qquad |\Im\sqrt{\lambda}| \neq 0. \tag{2.197}$$

which decays exponentially at infinity in the complex λ–plane, and therefore yields no contribution when integrated over a circular contour at $\mid\lambda\mid\to\infty$.

VI.3 Relation Between the Spectral (Characteristic) Green's Function and the Eigenvalue Problems

In this section, it is shown how the complete orthonormal set of eigenfunctions in Section VI.1 can be constructed from knowledge of the spectral Green's function in Section VI.2. In fact, it will become apparent that the Green's function route furnishes a far more general approach to the representation of wavefields.

To establish the Green's function–eigenfunction connection in qualitative physical terms, it is recalled that the Green's function represents the field due to a localized source, with the parameter λ in (2.165) proportional to the square of the spatial wavenumber (i.e, the squared frequency of the spatial oscillations). If the spatial frequency of the source is varied between 0 and ∞ in a lossless environment, the Green's function will exhibit amplitude singularities at each spatial frequency which corresponds to an eigenvalue λ_α; since λ_α identifies a *source-free* solution, the *driven* response at λ_α is unbounded. Therefore, the *totality* of singularities in the Green's function generates the *complete eigenfunction set.*

We begin by *assuming* that the set of eigenfunctions $\{f_\alpha\}$ is complete. The Green's function $g(u, u'; \lambda)$ may therefore be expanded in a series of eigenfunctions with coefficients g_α as

$$g(u, u'; \lambda) = \sum_\alpha g_\alpha(u'; \lambda) f_\alpha(u). \tag{2.198}$$

Applying the operator $L(u)$ in (2.128) to both sides of (2.198), and using (2.154) and (2.165), one obtains

$$-\sum_\alpha w(u) f_\alpha(u) f_\alpha^*(u') = \sum_\alpha g_\alpha(u'; \lambda) \left[\frac{d}{du} p(u) \frac{d}{du} - q(u) + \lambda w(u) \right] f_\alpha(u) . \tag{2.199}$$

Via (2.127), this can be written as

$$-\sum_\alpha w(u) f_\alpha(u) f_\alpha^*(u') = \sum_\alpha g_\alpha(u'; \lambda)(\lambda - \lambda_\alpha) w(u) f_\alpha(u) . \tag{2.200}$$

Equating the coefficients of the orthogonal functions $f_\alpha(u)$ on both sides of (2.200) yields

$$g_\alpha(u'; \lambda) = -\frac{f_\alpha^*(u')}{(\lambda - \lambda_\alpha)}. \tag{2.201}$$

Substitution of these coefficients into (2.199) gives

$$g(u, u'; \lambda) = -\sum_\alpha \frac{f_\alpha(u) f_\alpha^*(u')}{(\lambda - \lambda_\alpha)}, \tag{2.202}$$

which is an expression for the Green's function $g(u, u'; \lambda)$ in terms of the eigenfunctions of the homogeneous problem defined by (2.127).

The result in (2.202) can be used to derive a generalized completeness relation. Both sides of (2.202) are integrated in the complex-λ plane over a contour C which encloses in the counterclockwise sense all the pole singularities at the eigenvalues λ_α. The contour C is deformed into the contour C' consisting of small semicircles C_α centered at the poles λ_α and of line segments C'' which approach the real axis, as shown in Figure 2.11. The contributions to the integral due to the oppositely directed C''-segments along the real axis cancel, and each pair of semicircular arcs C_α contributes a residue at the corresponding pole as the radius of the semicircles approaches zero. Therefore, by the residue theorem, the line integral of g is

$$\begin{aligned}\frac{1}{2\pi j}\oint_C g(u,u';\lambda)\,d\lambda &= \frac{1}{2\pi j}\oint_{C'} g(u,u';\lambda)\,d\lambda\\ &= -\sum_\alpha f_\alpha(u)f_\alpha^*(u')\left(\frac{1}{2\pi j}\oint_{C_\alpha}\frac{d\lambda}{\lambda-\lambda_\alpha}\right)\\ &= \sum_\alpha f_\alpha(u)f_\alpha^*(u')\\ &= \frac{\delta(u-u')}{w(u')}.\end{aligned} \tag{2.203}$$

Equation (2.203) establishes the Green's function-eigenfunction connection in the completeness relation, which now takes the form

$$\frac{\delta(u-u')}{w(u')} = \frac{1}{2\pi j}\oint_C g(u,u';\lambda)\,d\lambda. \tag{2.204}$$

The contour C in Figure 2.11 can be terminated *anywhere* at $|\lambda|\to\infty$ because, as shown in SectionVI.2, g converges exponentially at $|\lambda|\to\infty$, so that contour segments at infinity do not contribute to the integral. The contour C must, however, have all of the singularities of g on one side. Because of this resolving connection with the eigenvalue problem, the spectral Green's function is also referred to as the characteristic (resolvent) Green's function. Although demonstrated here only for discrete eigenspectra (poles λ_α in the complex λ–plane), the characteristic Green's function procedure in (2.204) remains valid for continuous eigenspectra (typically in unbounded regions) which give rise to branch points in the then multi–sheeted complex λ–plane.

The importance of (2.204) resides in the fact that $g(u,u';\lambda)$ can be evaluated *directly* as in (2.187) of Section VI.2. Thus, (2.204) furnishes a generalized completeness relation for representing an arbitrary function $F(u)$. Such a representation is obtained, as before, by multiplying both sides of (2.204) by $w(u')F(u')$ and integrating over u' between the limits u_1 and u_2, giving

$$F(u) = \frac{1}{2\pi j}\oint_C d\lambda\, g(u,u';\lambda)\left\{\int_{u_1}^{u_2} du'\, w(u')F(u')\right\}. \tag{2.205}$$

VII Radiation and Edge Condition

VII.1 Radiation Condition

For an unbounded region it is necessary to specify the field behavior on a surface at infinity. By assuming that all sources are contained in a finite region, only

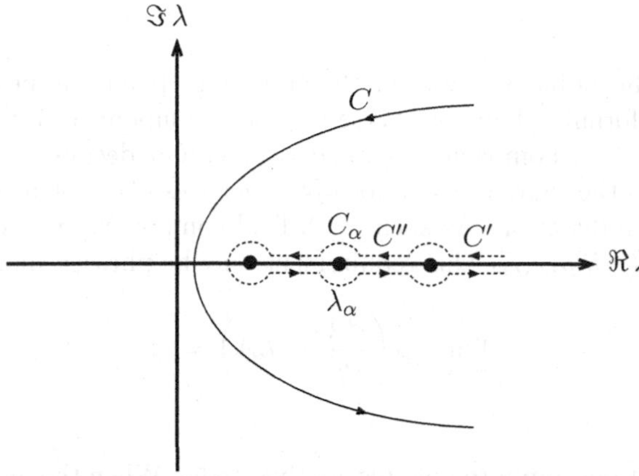

Fig. 2.11. Integration contours in the complex λ-plane.

outgoing waves can be present at large distances from the sources. In other words, the field behavior at large distances from the sources must meet the physical requirement that energy travel away from the source region. This requirement is the Sommerfeld *"radiation condition"* and constitutes a boundary condition on the surface at infinity. It assumes different expressions when dealing with 2D– or 3D–regions.

3D region.

Let A denote any field component transverse to the radial distance r. The transverse field of a spherically diverging wave in a homogeneous isotropic medium decays as $1/r$ at large distances r from the source region; locally the spherical wave behaves like a plane wave traveling in the *outward* r direction. As such (for an implied $e^{j\omega t}$ time dependence) each field component transverse to r must behave like $\exp(-jkr)/r$, where $k = \omega/c$ is the free–space wavenumber and c is the speed of light in vacuum. This requirement may be phrased mathematically as

$$\lim_{r\to\infty} r\left(\frac{\partial A}{\partial r} + jkA\right) = 0\,. \tag{2.206}$$

Observe that the above boundary condition is not self–adjoint in the Hermitian sense. The adjoint boundary condition would be

$$\lim_{r\to\infty} r\left(\frac{\partial A}{\partial r} - jkA\right) = 0\,. \tag{2.207}$$

corresponding to waves impinging from infinity.

2D region.

Let ρ denote the radial variable in the transverse plane, perpendicular to the direction of uniformity. The transverse to ρ field component A in a cylindrically diverging wave in a homogeneous isotropic medium decays as $1/\sqrt{\rho}$ at large distances ρ from the source region; locally A behaves like a plane wave travelling in the outward ρ direction. As such, each field component transverse to ρ must behave like $\exp(-jk\rho)/\sqrt{\rho}$ This requirement may be phrased mathematically as

$$\lim_{\rho\to\infty} \sqrt{\rho}\left(\frac{\partial A}{\partial \rho} + jkA\right) = 0\,. \tag{2.208}$$

The above equations apply to non–dissipative media. When the media are slightly lossy one may use the simpler requirement that all fields excited by sources in a finite region should vanish at infinity (i.e. k has a small *negative* imaginary part).

VII.2 Edge Condition in Two Dimensions

It is well recognized that, in many cases, boundary and radiation conditions alone are not sufficient to determine the solution uniquely [15, p.385], since it is possible to construct several different fields which satisfy these conditions [18]. As an example, let us consider a metallic wedge as shown in Figure 2.12, which we assume with no changes in the z–direction and separability in cylindrical coordinates. Assume a field E_0 which satisfies boundary and radiation conditions.

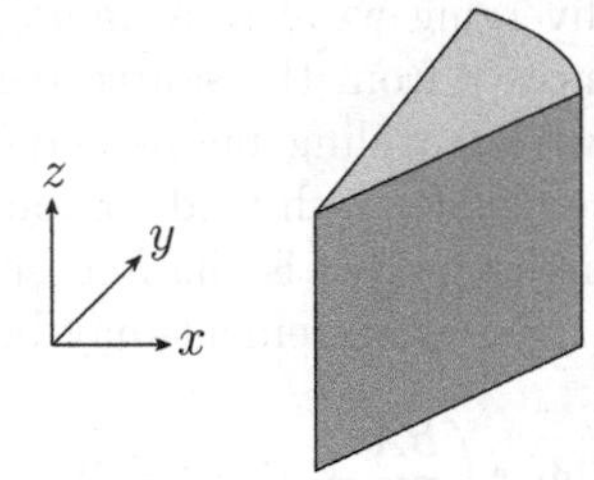

Fig. 2.12. Three-dimensional view of a perfectly conducting wedge extending from $\phi = 0$ to $\phi = \phi_2$ with no variations in the z–direction.

Now consider a field

$$E_z = E_0 + \bar{C} J_\nu(k\rho) \sin\left[\nu(\phi - \phi_2)\right] \tag{2.209}$$

which satisfies the Helmholtz equation (the scalar wave equation) for any value of $\bar{C}$, complies with the radiation condition and has the same boundary behavior as E_0 since $E_z = 0$ when $\phi = 0, \phi_2$. However, an infinite set of solutions can be generated by giving different values to $\bar{C}$ [14, pp.531-532]. Therefore, it is necessary to apply an additional constraint in order to achieve a unique solution i.e. an edge condition [18–20]. We start by noting that *the electromagnetic energy density must be integrable over any finite domain* even if this domain contains singularities of the electromagnetic field. Differently stated, *the electromagnetic energy in any finite domain must be finite.* The sum of the electric and magnetic energies in a small volume V surrounding the edge is [12, p. 24]

$$\frac{1}{2}\int_V \left(\varepsilon \boldsymbol{E}\cdot\boldsymbol{E}^* + \mu \boldsymbol{H}\cdot\boldsymbol{H}^*\right)\rho\, d\phi\, d\rho\, dz \tag{2.210}$$

In the vicinity of the edge the fields can be expressed as a power series in ρ; this series will have a dominant term ρ^μ where μ may be negative. Therefore, as ρ approaches zero the dominant term of the field components of (E, H) appearing in (2.210) behaves like $\rho^{2\mu}$, and the entire integrand behaves like $\rho^{2\mu+1}$. Integration over ρ yields $\rho^{2(\mu+1)}$ which is bounded for $\mu > -1$. The actual degree of singularity that a field experiences near the edge is dependent on the wedge configuration. It is also noted that the field singularity does not depend on frequency since in the proximity of the edge, spatial derivatives of the fields are much larger than time derivatives, so that the latter can be neglected in Maxwell's equations (quasi–static regime).
The exact knowledge of the type of field singularity near the edge is of considerable importance for numerical applications. The reader may find more information on cases of practical importance in reference [21].

VIII Reciprocity and Field Equivalence Principles

VIII.1 Reaction in Electromagnetic Theory

The *reaction concept* in electromagnetic theory has been introduced in [22] in order to find a fundamental observable representing measurements which can be performed practically. For example, if we want to measure the field radiated by some source of electromagnetic energy, we may use an antenna probe and observe the signal received at terminals at the point of observation. However, the latter measurement does not provide the field *just* at the observation point, but it measures the effect of the field over a small, but finite, region. To take this fact into account, it is convenient to define the *reaction*, i.e. the coupling between the field that we want to measure and the antenna that we are using.
Consider a monochromatic source of electromagnetic field, denoted by a, consisting of electric and magnetic currents $\boldsymbol{J}_a$ and $\boldsymbol{M}_a$, respectively, and producing the field $\boldsymbol{E}_a$, $\boldsymbol{H}_a$. Similarly, consider also a source b of electric and magnetic currents $\boldsymbol{J}_b$ and $\boldsymbol{M}_b$, generating the field $\boldsymbol{E}_b$, $\boldsymbol{H}_b$. The interaction of source a with field b may be characterized by the complex number $\langle a, b\rangle$, defined as [4]

$$\langle a, b \rangle = \int_V (\boldsymbol{J}_a \cdot \boldsymbol{E}_b - \boldsymbol{M}_a \cdot \boldsymbol{H}_b)\, dV\,, \tag{2.211}$$

where the first entry, a, is associated with the source (or probe), and the second entry, b, is associated with the observed field. The integration is extended over the volume V, i.e. the region containing the source a, which may contain both volume current densities and surface current densities. Note that, for an ideal electric field probe, $\boldsymbol{J}_a$ is a delta function which measures the field just at the observation point. As noted in the previous paragraph, also for electromagnetic field quantities, the reaction is different from *complex power* since there is no complex–conjugate. Moreover, let Σ represent any scalar and Σa be the source a increased in strength by the factor Σ, then

$$\langle \Sigma a, b \rangle = \Sigma \langle a, b \rangle\,. \tag{2.212}$$

By considering another source c, radiating at the same frequency as a and b, we have

$$\langle a, (b + c) \rangle = \langle a, b \rangle + \langle a, c \rangle\,. \tag{2.213}$$

VIII.2 Lorentz Reciprocity Theorem

Having discussed the reaction concept we proceed to the Lorentz reciprocity theorem. A simple interpretation of this theorem is that, in isotropic media, the response of a system to a source is unchanged when source and detector are interchanged [13]. In order to establish this theorem let us consider the two monochromatic sources a, b and the field produced thereby. In each case Maxwell's equations are:

$$\nabla \times \boldsymbol{E}_a = -j\omega\mu \boldsymbol{H}_a - \boldsymbol{M}_a\,, \tag{2.214a}$$

$$\nabla \times \boldsymbol{H}_a = j\omega\varepsilon \boldsymbol{E}_a + \boldsymbol{J}_a \tag{2.214b}$$

and

$$\nabla \times \boldsymbol{E}_b = -j\omega\mu \boldsymbol{H}_b - \boldsymbol{M}_b\,, \tag{2.215a}$$

$$\nabla \times \boldsymbol{H}_b = j\omega\varepsilon \boldsymbol{E}_b + \boldsymbol{J}_b\,. \tag{2.215b}$$

Performing dot product multiplication of (2.214b) by $\boldsymbol{E}_b$ and of (2.215a) by $\boldsymbol{H}_a$, and subtracting one from the other we obtain

$$\nabla \cdot (\boldsymbol{E}_b \times \boldsymbol{H}_a) = -j\omega\varepsilon \boldsymbol{E}_a \cdot \boldsymbol{E}_b - \boldsymbol{J}_a \cdot \boldsymbol{E}_b - j\omega\mu \boldsymbol{H}_a \cdot \boldsymbol{H}_b - \boldsymbol{M}_b \cdot \boldsymbol{H}_b\,, \tag{2.216}$$

where use is made of the identity

$$\nabla \cdot (\boldsymbol{U} \times \boldsymbol{V}) = \boldsymbol{V} \cdot \nabla \times \boldsymbol{U} - \boldsymbol{U} \cdot \nabla \times \boldsymbol{V}\,. \tag{2.217}$$

Similarly, performing dot product multiplication of (2.215b) by $\boldsymbol{E}_a$, and of (2.214a) by $\boldsymbol{H}_b$, and subtracting one from the other, we obtain:

$$\nabla \cdot (\boldsymbol{E}_a \times \boldsymbol{H}_b) = -j\omega\varepsilon \boldsymbol{E}_a \cdot \boldsymbol{E}_b - \boldsymbol{J}_b \cdot \boldsymbol{E}_a - j\omega\mu \boldsymbol{H}_a \cdot \boldsymbol{H}_b - \boldsymbol{M}_a \cdot \boldsymbol{H}_b . \quad (2.218)$$

Finally, by subtracting (2.216) from (2.218), integrating throughout a source–free region, and applying the divergence theorem we arrive at

$$\oint_S (\boldsymbol{E}_a \times \boldsymbol{H}_b - \boldsymbol{E}_b \times \boldsymbol{H}_a)\, d\boldsymbol{S} = \langle a, b\rangle - \langle b, a\rangle . \quad (2.219)$$

By *definition*, isotropic media are reciprocal when

$$\oint_S (\boldsymbol{E}_a \times \boldsymbol{H}_b - \boldsymbol{E}_b \times \boldsymbol{H}_a)\, d\boldsymbol{S} = 0 . \quad (2.220)$$

In this case, the Lorentz reciprocity theorem can be stated as

$$\langle a, b\rangle - \langle b, a\rangle = 0. \quad (2.221)$$

The surface integral on the left side of (2.219) vanishes also when the surface S encloses all the sources. In fact, in this case we can consider the complementary source–free volume bounded by S and the surface S_∞ of a sphere with infinite radius. When the fields satisfy the radiation condition the integrand of the left side of (2.219) vanishes on S_∞ and (2.220) applies as well.

The Lorentz theorem has a variety of useful applications. It allows one to derive stationary formulas in variational problems in a direct manner. It is also suitable for proving simple assertions, such as the fact that an electric current sheet impressed on the surface of a perfect conductor does not radiate [4]. This is a trivial result when the surface of the conductor is planar, since image theory shows that no field is produced. In fact, by replacing the metallic plane by an image source, i.e. by an impressed current directed in the opposite direction, the two impressed currents annihilate, producing zero field. When the surface is not planar, application of reciprocity demonstrates the above assertion in the following way. With reference to Figure 2.13 let us consider source a on the perfect electric conductor. In order to measure the field $\boldsymbol{E}_a$, $\boldsymbol{H}_a$ produced by this source, let us place a probe (source b) at the observation point and evaluate the reaction of source b on the field a, i.e. $\langle b, a\rangle$. By the reciprocity theorem, the effect of source b on the field a is equal to the effect of source a on the field b, i.e.

$$\langle b, a\rangle = \langle a, b\rangle . \quad (2.222)$$

However, the tangential component of the electric field produced by b is zero on the metallic surface where $\boldsymbol{J}_a$ is present, thus

$$\langle a, b\rangle = 0 . \quad (2.223)$$

In view of the arbitrariness of source b it is proved that the impressed electric current sheets $\boldsymbol{J}_a$ on the surface of the perfect electric conductor do not produce any field.

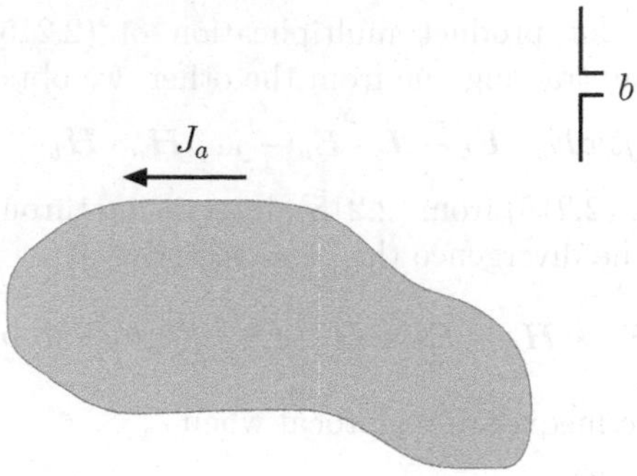

Fig. 2.13. The impressed electric current sheets $\boldsymbol{J}_a$ on the surface of a perfect electric conductor do not produce any field, as measured through probe b.

Huygens' Principle

The propagation of electromagnetic fields can be visualized according to Christian Huygens as wavefronts comprising a number of secondary sources or radiators, each generating new spherical wavelets. According to Huygens' principle the envelope of these wavelets forms a wavefront which in turn consists of new sources giving rise to a new generation of spherical wavelets. This in turn means that the field solution in a region is completely determined by the tangential fields specified over the surface enclosing the region. This principle can be rigorously stated in mathematical terms, as shown next. To this end we need to recall scalar and vector Green's theorems [3] which, as noted in [13, p.120], are mathematical statements of reciprocity (symmetrical in two functions). The difference between the Lorentz reciprocity theorem and Green's theorem is that no physical interpretation is ascribed to the latter.

Scalar Green's Theorem

Consider a closed regular surface S bounding a volume V where the two scalar functions $\bar{\phi}$ and $\bar{\psi}$, continuous together with their first and second derivatives throughout V and on the surface S, are defined. Applying the divergence theorem to the vector $\bar{\psi}\nabla\bar{\phi}$ yields

$$\int_V \nabla \cdot (\bar{\psi}\nabla\bar{\phi})\, dV = \int_S (\bar{\psi}\nabla\bar{\phi}) \cdot \boldsymbol{n}\, dS\,. \tag{2.224}$$

The divergence on the left–hand side may be expanded as

$$\nabla \cdot (\bar{\psi}\nabla\bar{\phi}) = \nabla\bar{\psi} \cdot \nabla\bar{\phi} + \bar{\psi}\nabla \cdot \nabla\bar{\phi} = \nabla\bar{\psi} \cdot \nabla\bar{\phi} + \bar{\psi}\nabla^2\bar{\phi}\,. \tag{2.225}$$

while on the right–hand side, we may replace the normal component of the gradient by the normal derivative, i.e.

$$\nabla\bar{\phi}\cdot\boldsymbol{n}=\frac{\partial\bar{\phi}}{\partial n}\,. \tag{2.226}$$

Upon substituting (2.225) and (2.226) into (2.224) we obtain Green's *first* identity

$$\int_V \nabla\bar{\psi}\cdot\nabla\bar{\phi}\,dV+\int_V \bar{\psi}\nabla^2\bar{\phi}\,dV=\int_S \bar{\psi}\frac{\partial\bar{\phi}}{\partial n}\,dS\,. \tag{2.227}$$

This identity holds also when interchanging the roles of the functions $\bar{\phi},\bar{\psi}$; by so doing we obtain

$$\int_V \nabla\bar{\psi}\cdot\nabla\bar{\phi}\,dV+\int_V \bar{\phi}\nabla^2\bar{\psi}\,dV=\int_S \bar{\phi}\frac{\partial\bar{\psi}}{\partial n}\,dS\,. \tag{2.228}$$

Subtracting (2.228) from (2.227) we get another important identity, Green's *second* identity, namely,

$$\int_V \left(\bar{\psi}\nabla^2\bar{\phi}-\bar{\phi}\nabla^2\bar{\psi}\right)dV=\int_S \bar{\psi}\frac{\partial\bar{\phi}}{\partial n}\,dS-\int_S \bar{\phi}\frac{\partial\bar{\psi}}{\partial n}\,dS\,, \tag{2.229}$$

which is frequently referred to as *Green's theorem.*

Vector Green's Theorem

Let us return to the surface S and volume V as defined in the previous paragraph, but consider two vector functions $\boldsymbol{U}$ and $\boldsymbol{V}$ which, together with their first and second derivatives, are continuous throughout V and on the surface S. Then, replacing the gradient by the curl, i.e. ∇ by $\nabla\times$, and ∇^2 by $\nabla\times\nabla\times$, we have the building blocks for the vector analogue of the scalar Green's theorem. Applying the divergence theorem to the vector $\boldsymbol{U}\times\nabla\times\boldsymbol{V}$,

$$\int_V \nabla\cdot(\boldsymbol{U}\times\nabla\times\boldsymbol{V})\,dV=\int_S (\boldsymbol{U}\times\nabla\times\boldsymbol{V})\cdot\boldsymbol{n}dS \tag{2.230}$$

and expanding the divergence on the left hand side we get

$$\begin{gathered}\nabla\cdot(\boldsymbol{U}\times\nabla\times\boldsymbol{V})=\\ \nabla_P\cdot(\boldsymbol{U}\times\nabla\times\boldsymbol{V})+\nabla_Q\cdot(\boldsymbol{U}\times\nabla\times\boldsymbol{V})=\\ \nabla\times\boldsymbol{U}\cdot\nabla\times\boldsymbol{V}-\boldsymbol{U}\cdot\nabla\times\nabla\times\boldsymbol{V}\end{gathered} \tag{2.231}$$

which, by substitution into (2.230), provides the vector analogue of Green's first identity,

$$\int_V (\nabla\times\boldsymbol{U}\cdot\nabla\times\boldsymbol{V})-(\boldsymbol{U}\cdot\nabla\times\nabla\times\boldsymbol{V})\,dV=\int_S (\boldsymbol{U}\times\nabla\times\boldsymbol{V})\cdot\boldsymbol{n}\,dS\,. \tag{2.232}$$

Another form of the vector first identity may be obtained by interchanging $\boldsymbol{U}$ and $\boldsymbol{V}$,

$$\int_V (\nabla \times \boldsymbol{V} \cdot \nabla \times \boldsymbol{U}) - (\boldsymbol{V} \cdot \nabla \times \nabla \times \boldsymbol{U})\, dV = \int_S (\boldsymbol{V} \times \nabla \times \boldsymbol{U}) \cdot \boldsymbol{n} dS\,. \quad (2.233)$$

By subtracting (2.233) from (2.232) we get the vector analogue of Green's *second* identity,

$$\int_V (\boldsymbol{V} \cdot \nabla \times \nabla \times \boldsymbol{U} - \boldsymbol{U} \cdot \nabla \times \nabla \times \boldsymbol{V})\, dV =$$

$$\int_S (\boldsymbol{U} \times \nabla \times \boldsymbol{V} \ - \boldsymbol{V} \times \nabla \times \boldsymbol{U}) \cdot \boldsymbol{n} dS\,. \quad (2.234)$$

The dyadic form of *Huygens' principle* is obtained on replacing the vector $\boldsymbol{V}$ in (2.234) by the scalar product of Green's dyad $\mathscr{G}$ with a vector $\boldsymbol{U}$, i.e. $\boldsymbol{V} = \mathscr{G} \cdot \boldsymbol{U}$.

Mathematical Formulation of Huygens' Principle

The equivalence principle is rigorously proved by introducing the mathematical formulation of Huygens' principle [3, 12, 23].
Consider a volume V, containing all sources, bounded by a smooth surface S. The electric field in V is a solution of the source–free vector wave equation

$$\nabla \times \nabla \times \boldsymbol{E} - k^2 \boldsymbol{E} = 0. \quad (2.235)$$

Consider also the dyadic Green's function $\mathscr{G}_e$ which, in turn, is a solution of

$$\nabla \times \nabla \times \mathscr{G}_e - k^2 \mathscr{G}_e = \mathscr{I} \delta(\boldsymbol{r} - \boldsymbol{r}')\,. \quad (2.236)$$

Both $\boldsymbol{E}$ and $\boldsymbol{G}_e$ satisfy the electric field boundary conditions on S as well as the radiation condition at infinity. Here $\mathscr{I}$ is the identity dyadic and $\delta(\boldsymbol{r} - \boldsymbol{r}')$ is the Dirac delta function. Forming the scalar products

$$\boldsymbol{E} \cdot \nabla \times \nabla \times \mathscr{G}_e - \nabla \times \nabla \times \boldsymbol{E} \cdot \mathscr{G}_e \quad (2.237)$$

and then applying Green's vector second identity in (2.234), yields

$$\int_V (\boldsymbol{E} \cdot \nabla \times \nabla \times \mathscr{G}_e - \nabla \times \nabla \times \boldsymbol{E} \cdot \mathscr{G}_e)\, dV =$$

$$\int_S (\mathscr{G}_e \times \nabla \times \boldsymbol{E} - \boldsymbol{E} \times \nabla \times \mathscr{G}_e) \cdot \boldsymbol{n} dS\,, \quad (2.238)$$

where the integral over the sphere at infinity has been set to zero because both $\boldsymbol{E}$ and $\mathscr{G}_e$ satisfy the radiation condition. Hence, using (2.236) we have

$$\int_S (\boldsymbol{n} \times \boldsymbol{E} \cdot \nabla \times \mathscr{G}_e + \boldsymbol{n} \times \nabla \times \boldsymbol{E} \cdot \mathscr{G}_e)\, dS = \begin{cases} \boldsymbol{E}(\boldsymbol{r}') & \boldsymbol{r}' \text{ in } V \\ \boldsymbol{0} & \boldsymbol{r}' \text{ in } V_1 \end{cases}. \tag{2.239}$$

Using Maxwell's curl equation,

$$\nabla \times \boldsymbol{E} = -j\omega\mu \boldsymbol{H}, \tag{2.240}$$

(2.239) may be written in terms of the currents flowing on S as

$$\boldsymbol{E}(\boldsymbol{r}') = \int_S \boldsymbol{n} \times \boldsymbol{E} \cdot \nabla \times \mathscr{G}_e dS - j\omega\mu \int_S \boldsymbol{n} \times \boldsymbol{H} \cdot \mathscr{G}_e dS\,. \tag{2.241}$$

The formula in (2.241) provides the electric field at each point of V in terms of the boundary fields on S, and constitutes the mathematical version of Huygens' principle [12, p.135], [23]). By following the same steps, or by using duality, it is possible to derive a formula analogous to (2.241) for the magnetic field, i.e.

$$\boldsymbol{H}(\boldsymbol{r}') = \int_S \boldsymbol{n} \times \boldsymbol{H} \cdot \nabla \times \mathscr{G}_m dS + j\omega\varepsilon \int_S \boldsymbol{n} \times \boldsymbol{E} \cdot \mathscr{G}_m dS\,, \tag{2.242}$$

where the magnetic field dyadic Green's function $\mathscr{G}_m$ satisfies (2.236) with magnetic field boundary conditions and the radiation condition.
By recalling the equivalence theorem, it follows that specification of the tangential components of the $\boldsymbol{E}, \boldsymbol{H}$ fields on S is the same as the specification of equivalent electric and magnetic currents $\boldsymbol{J}$ and $\boldsymbol{M}$. It is useful to write (2.241) and (2.242) operationally in the following way

$$\boldsymbol{E} = \hat{Z}(\boldsymbol{J}) + \hat{T}_e(\boldsymbol{M})\,, \tag{2.243a}$$

$$\boldsymbol{H} = \hat{T}_m(\boldsymbol{J}) + \hat{Y}(\boldsymbol{M})\,, \tag{2.243b}$$

which express the electromagnetic field (as obtained from the field on S) in terms of operators identified from (2.241) and (2.242). It can be proved by inserting (2.243b) into (2.14a) and (2.14b), and noting the arbitrariness of $\boldsymbol{J}$ and $\boldsymbol{M}$, that the above operators also satisfy the following equations

$$\nabla \times \hat{Z} = -j\omega\mu\hat{T}_m\,, \tag{2.244a}$$

$$\nabla \times \hat{Y} = j\omega\varepsilon\hat{T}_e\,, \tag{2.244b}$$

$$\nabla \times \hat{T}_e = -j\omega\mu\hat{Y} - \boldsymbol{M}\,, \tag{2.244c}$$

$$\nabla \times \hat{T}_m = j\omega\varepsilon\hat{Z} + \boldsymbol{J}\,, \tag{2.244d}$$

from which one obtains

$$(\nabla \times \nabla \times -k^2)\hat{Z} = -j\omega\mu\boldsymbol{J}\,, \tag{2.245a}$$

$$(\nabla \times \nabla \times -k^2)\hat{Y} = -j\omega\varepsilon\boldsymbol{M}\,, \tag{2.245b}$$

$$(\nabla \times \nabla \times -k^2)\hat{T}_e = -\nabla \times \boldsymbol{M}\,, \tag{2.245c}$$

$$(\nabla \times \nabla \times -k^2)\hat{T}_m = \nabla \times \boldsymbol{J}\,, \tag{2.245d}$$

The operators $\hat{Z}, \hat{T}_e$ satisfy the same boundary condition as the electric field, while the operators $\hat{Y}, \hat{T}_m$ satisfy the same boundary condition as the magnetic field.

An interesting circuit analogy of (2.243b) can be obtained by considering the equivalent sources $\boldsymbol{J}$ and $\boldsymbol{M}$ on the surface S_1 and the observation point $\boldsymbol{r}'$ on the surface S_2. In this case, (2.243b) corresponds to an $ABCD$ representation of the region of space between the two surfaces. In order to describe this region, we could also have chosen other representations, such as the Z (impedance) or the Y (admittance) representation. As an example, a Z representation is obtained by considering the two surfaces S_1, S_2 as *magnetic* walls. Accordingly only the electric currents, i.e. the magnetic fields, produce radiation away from the surfaces. By letting $\boldsymbol{E}_1, \boldsymbol{H}_1$ be the electric and magnetic fields on the surface S_1, and $\boldsymbol{E}_2, \boldsymbol{H}_2$ the electric and magnetic fields on the surface S_2, we may express the (impedance) relationship between electric and magnetic fields on these surfaces as

$$\boldsymbol{E}_1 = \hat{Z}_{11}(\boldsymbol{H}_1) + \hat{Z}_{12}(\boldsymbol{H}_2)\,, \tag{2.246a}$$

$$\boldsymbol{E}_2 = \hat{Z}_{21}(\boldsymbol{H}_1) + \hat{Z}_{22}(\boldsymbol{H}_2)\,. \tag{2.246b}$$

A similar relationship may be written for the admittance representation.

Finally note that, when the operator is expressed in a diagonalized form, i.e. when the region we are dealing with is coordinate and vector separable, we can pass from one representation, say the admittance representation, to another representation.

References

[1] J. C. Maxwell, *A Treatise on Electricity and Magnetism.* London: Clarendon press, 1891.

[2] R. S. Elliott, *Electromagnetics.* New York: IEEE Press, 1993.

[3] J. A. Stratton, *Electromagnetic Theory.* New York, NY: McGraw-Hill, 1941.

[4] J. A. Kong, *Electromagnetic Wave Theory.* Singapore: John Wiley & Sons, 1986.

[5] I. V. Lindell and A. Sihvola, "Perfect electromagnetic conductor," in *Proc. 9th International Conference on Electromagnetics in Advanced Applications*, Sept. 12 - 16 2005.

[6] P. Russer, "Electromagnetic properties and realisability of gyrator surfaces," in *ICEAA2007*, Torino, Italy, sept 2007, pp. 320–323.

[7] B. D. H. Tellegen, "A general network theorem with applications," *Philips Research Reports*, vol. 7, pp. 259–269, 1952.

[8] ——, "A general network theorem with applications," *Proc. Inst. Radio Engineers*, vol. 14, pp. 265–270, 1953.

[9] P. Penfield, R. Spence, and S. Duinker, *Tellegen's theorem and electrical networks.* Campbridge, Massachusetts: MIT Press, 1970.

[10] P. Russer, M. Mongiardo, and L. B. Felsen, "Electromagnetic field representations and computations in complex structures: network representations of the connection and subdomain circuits," *International Journal of Numerical Modeling: Electronic Networks, Devices and Fields*, vol. 15, pp. 127–145, 2002.

[11] P. Russer, *Electromagnetics, Microwave Circuit and Antenna Design for Communications Engineering*, 2nd ed. Boston: Artech House, 2006.

[12] R. E. Collin, *Field Theory of Guided Waves.* New York: IEEE Press, 1991.

[13] R. F. Harrington, *Time Harmonic Electromagnetic Fields.* New York: McGraw-Hill, 1961.

[14] D. S. Jones, *Acoustic and Electromagnetic Waves.* Oxford, England: Clarendon Press, 1986.

[15] J. V. Bladel, *Electromagnetic Fields.* New York: McGraw-Hill, 1964.

[16] L. B. Felsen and N. Marcuvitz, *Radiation and Scattering of Waves.* Englewood Cliffs, NJ: Prentice Hall, 1973, Piscataway, NJ: IEEE Press (classic reissue), 1994.

[17] P. Morse and H. Feshbach, *Methods of Theoretical Physics, Part 1.* New York: McGraw-Hill, 1953.

[18] C. J. Bouwkamp, "A note on singularities occurring at sharp edges in electromagnetic diffraction theory," *Physica*, vol. 12, pp. 467–474, 1946.

[19] J. Meixner, "Die Kantenbedingung in der Theorie der Beugung elektromagnetischer Wellen an vollkommen leitenden ebenen Schirmen," *Ann. Phys.*, vol. 6, pp. 1–9, 1949.

[20] ——, "The behaviour of electromagnetic fields at edges," *IEEE Trans. Antennas Propagat.*, vol. 20, no. 4, pp. 442–446, Jul. 1972.

[21] T. Rozzi and M. Mongiardo, *Open Electromagnetic Waveguides.* London: IEE, 1997.

[22] V. H. Rumsey, "Reaction concept in electromagnetic theory," *Phys. Rev., ser. 2*, vol. 94, no. 6, pp. 1483–1491, 1954.

[23] C. T. Tai, *Dyadics Green's Functions in Electromagnetic Theory.* Scranton, PA: Intext Educational Publishers, 1971.

3

Wave–Guiding Configurations

I Introduction

As noted in Section II of Chapter 1, we are considering an architecture that decomposes a complex conglomerate into simpler interactive subdomains. Subdomains (SD) can be treated in a variety of ways ranging from purely numerical methods, such as finite element or finite difference methods, to analytic approaches based on constructible problem–matched Green's functions (GF) [1–3]. A model SD Green's function that is *well–matched* to the actual SD problem configuration can serve as an efficient *background kernel* in the integral equation formulation for the *actual* SD–GF.

When the actual problem is so irregular as to render GF–matching impractical, the homogeneous free–space GF may be the only (but least efficient) analytic option, apart from purely numerical methods [4–6]. Depending on the SD problem parameters, the algorithmic efficiency can sometimes be enhanced by quasi–static or high frequency asymptotic extractions, re summing of series, etc., as appropriate [7]. Analytic techniques, when feasible, provide physical insight, and yield more efficient field representations as well as computations.

The "cleanest" model Green's functions (GFs) are based on configurations that render the vector field equations, with boundary conditions, at least partially coordinate–separable [8–10]. In the construction of coordinate–separable Green's functions (GFs), alternative representations, which impact the rapidity of convergence and wave–physical interpretation of the associated algorithms, play a critical role [11]. Such alternatives and the relationships connecting them are best explored in the complex wavenumber–frequency spectral domain; the various separable options differ according to the "propagation coordinate" that is selected, and also according to the choice of boundary conditions at the interfaces (ports) between adjacent subdomains. These interface conditions can be phrased in terms of oscillatory wave, traveling wave and hybrid combinations, with corresponding choices of "primary" and "secondary" fields, which constitute the excitation of the SD and the SD response to that excitation, respectively. These alternative fields on the SD interconnects, in turn, give rise to corresponding alternative network representations.

To implement the coordinate-separable Green's function analysis noted above, it is necessary to structure the source–excited full Maxwell field equations accordingly. First, one identifies "uniform waveguide" regions with the preferred rectilinear coordinate z, along which "transmission" (propagation) is assumed to take place [12]. The non–changing cross–sections S transverse to z may be bounded by an as yet arbitrarily shaped perfectly electrically conducting (PEC) contour $\boldsymbol{\rho}(s)$ which may however extend to infinity (see Figure 3.1). To this end, Maxwell's equations are structured so as to separate the transverse (to z) field components from the longitudinal (z) components. The transverse fields are then expanded into a complete set of transverse orthogonal vector eigenfunctions (vector modes) which individually satisfy the boundary conditions on s.

The scalar amplitude of each source–excited vector mode field is a function of the longitudinal coordinate z which satisfies "Transmission Line Equations". The reduction from the full Maxwell vector field equations to the scalar modal transmission line equations, and to the ensuing scalarization of the dyadic Green's function [13] in a modal basis, is presented sequentially below.

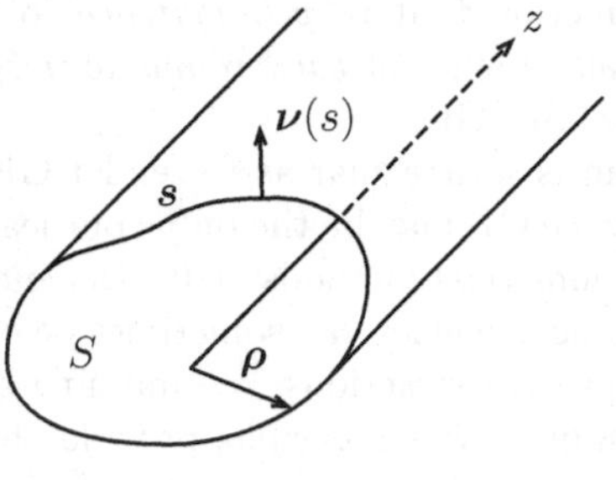

Fig. 3.1. Non–changing cross section S of a uniform waveguide region bounded by the contour $\boldsymbol{\rho}(s)$ where $\boldsymbol{\rho}$ is the transverse radial vector coordinate and $\boldsymbol{\nu}(s)$ is the outward normal unit vector to $\boldsymbol{\rho}(s)$ lying in the plane S, with $\boldsymbol{s}$ representing the length coordinate along the boundary.

II The Transverse Field Equations

II.1 Source–Free Case

We start with the source–free time–harmonic ($\exp(\jmath\omega t)$) Maxwell equations,

$$\nabla \times \boldsymbol{E} = -\jmath k\eta \boldsymbol{H}\,, \tag{3.1a}$$

$$\nabla \times \boldsymbol{H} = \jmath k\bar{\zeta} \boldsymbol{E}\,, \tag{3.1b}$$

where the wave impedance η and admittance $\bar{\zeta}$ of the medium are defined in terms of the medium permeability μ and permittivity ϵ as:

$$\eta = \frac{1}{\bar{\zeta}} = \sqrt{\frac{\mu}{\varepsilon}}. \tag{3.2}$$

In the following $k = \omega(\mu\varepsilon)^{1/2}$ is the wavenumber in the region and $\mathscr{I}$ is a unit dyadic such that $\mathscr{I} \cdot \boldsymbol{U} = \boldsymbol{U} \cdot \mathscr{I} = \boldsymbol{U}$.
We derive an invariant transverse vector notation for the Maxwell's field equations in a homogeneous and source–free medium by elimination of the field components along the transmission direction, the z-axis. Introducing the transverse gradient operator, $\nabla_t = \nabla - \boldsymbol{z}_0\frac{\partial}{\partial z}$, Maxwell's curl equations can be written as

$$\left(\nabla_t + \boldsymbol{z}_0\frac{\partial}{\partial z}\right) \times (\boldsymbol{E}_t + \boldsymbol{z}_0 E_z) = -jk\eta\,(\boldsymbol{H}_t + \boldsymbol{z}_0 H_z)\,,$$

$$\nabla_t \times \boldsymbol{E}_t + \nabla_t \times \boldsymbol{z}_0 E_z + \boldsymbol{z}_0 \times \frac{\partial \boldsymbol{E}_t}{\partial z} = -jk\eta\,(\boldsymbol{H}_t + \boldsymbol{z}_0 H_z)\,, \tag{3.3}$$

$$\left(\nabla_t + \boldsymbol{z}_0\frac{\partial}{\partial z}\right) \times (\boldsymbol{H}_t + \boldsymbol{z}_0 H_z) = jk\bar{\zeta}\,(\boldsymbol{E}_t + \boldsymbol{z}_0 E_z)\,,$$

$$\nabla_t \times \boldsymbol{H}_t + \nabla_t \times \boldsymbol{z}_0 H_z + \boldsymbol{z}_0 \times \frac{\partial \boldsymbol{H}_t}{\partial z} = jk\bar{\zeta}\,(\boldsymbol{E}_t + \boldsymbol{z}_0 E_z)\,. \tag{3.4}$$

By equating the terms in (3.3) and (3.4) according to their vector dependence we obtain

$$\nabla_t \times \boldsymbol{E}_t = -jk\eta \boldsymbol{z}_0 H_z\,, \tag{3.5a}$$

$$\nabla_t \times \boldsymbol{z}_0 E_z + \boldsymbol{z}_0 \times \frac{\partial \boldsymbol{E}_t}{\partial z} = -jk\eta \boldsymbol{H}_t\,, \tag{3.5b}$$

$$\nabla_t \times \boldsymbol{H}_t = jk\bar{\zeta}\boldsymbol{z}_0 E_z\,, \tag{3.5c}$$

$$\nabla_t \times \boldsymbol{z}_0 H_z + \boldsymbol{z}_0 \times \frac{\partial \boldsymbol{H}_t}{\partial z} = jk\bar{\zeta}\boldsymbol{E}_t\,. \tag{3.5d}$$

Applying the transverse curl operator to (3.5c) and substituting into (3.5b) yields

$$\boldsymbol{z}_0 \times \frac{\partial \boldsymbol{E}_t}{\partial z} = -\frac{1}{jk\bar{\zeta}}\nabla_t \times \nabla_t \times \boldsymbol{H}_t - jk\eta \boldsymbol{H}_t$$

and after performing the vector product with $-\boldsymbol{z}_0$ and applying the vector identity

$$\boldsymbol{z}_0 \times \nabla_t \times \nabla_t \times \boldsymbol{U} = \nabla_t \nabla_t (\boldsymbol{U} \times \boldsymbol{z}_0) \tag{3.6}$$

we obtain

$$\frac{\partial \boldsymbol{E}_t}{\partial z} = -jk\eta\left(\mathscr{I} + \frac{1}{k^2}\nabla_t\nabla_t\right) \cdot (\boldsymbol{H}_t \times \boldsymbol{z}_0) \, . \tag{3.7}$$

Similarly from (3.5a) and (3.5d), one obtains

$$\frac{\partial \boldsymbol{H}_t}{\partial z} = -jk\bar{\zeta}\left(\mathscr{I} + \frac{1}{k^2}\nabla_t\nabla_t\right) \cdot (\boldsymbol{z}_0 \times \boldsymbol{E}_t) \, . \tag{3.8}$$

By forming the scalar product of $\boldsymbol{z}_0$ with (3.5a) and (3.5c) respectively, we obtain the following longitudinal components of the electric and magnetic fields,

$$E_z = \frac{1}{jk\bar{\zeta}}\nabla_t \cdot (\boldsymbol{H}_t \times \boldsymbol{z}_0) = \frac{1}{jk\bar{\zeta}}\boldsymbol{z}_0 \cdot (\nabla_t \times \boldsymbol{H}_t) \, , \tag{3.9a}$$

$$H_z = \frac{1}{jk\eta}\nabla_t \cdot (\boldsymbol{z}_0 \times \boldsymbol{E}_t) = -\frac{1}{jk\eta}\boldsymbol{z}_0 \cdot (\nabla_t \times \boldsymbol{E}_t) \, . \tag{3.9b}$$

II.2 Source–Excited Case

Using the decomposition in Section II.1 we now consider the steady-state vector fields excited by a specified electric current distribution $\boldsymbol{J}(\boldsymbol{r})$ and magnetic current distribution $\boldsymbol{M}(\boldsymbol{r})$ in the waveguide environment of Figure 3.1. The source-excited Maxwell equations are:

$$\nabla \times \boldsymbol{E}(\boldsymbol{r}) = -j\omega\mu\boldsymbol{H}(\boldsymbol{r}) - \boldsymbol{M}(\boldsymbol{r}) \, , \tag{3.10}$$

$$\nabla \times \boldsymbol{H}(\boldsymbol{r}) = j\omega\varepsilon\boldsymbol{E}(\boldsymbol{r}) + \boldsymbol{J}(\boldsymbol{r}) \, . \tag{3.11}$$

On the perfectly conducting boundary of the uniform waveguide (see Figure 3.1), the tangential component of the electric field must vanish, i.e.,

$$\boldsymbol{\nu} \times \boldsymbol{E} = 0 \, , \qquad \text{on } s \, . \tag{3.12}$$

The vanishing of the tangential component of $\boldsymbol{E}$ on s also implies the vanishing on s of the normal component of $\boldsymbol{H}$. For a region with infinite cross section, condition (3.12) is replaced by a *radiation condition* which requires that, for any source distribution contained in a finite region, the field solution at infinity comprises only "outgoing" waves (see Chapter 2 Section VII). The boundary conditions on the longitudinal (z) boundaries of the region are left open for the moment and will be taken into account in the subsequent solution of the transmission-line equations. For the present discussion, the scalar permittivity ε and permeability μ of the waveguide medium may both be z dependent. To effect the transverse–longitudinal decomposition as in Section II.1, we take vector and scalar products of (3.11) with the longitudinal unit vector $\boldsymbol{z}_0$,

$$\begin{aligned} j\omega\mu\boldsymbol{H} \times \boldsymbol{z}_0 + \boldsymbol{M} \times \boldsymbol{z}_0 &= \boldsymbol{z}_0 \times (\nabla \times \boldsymbol{E}) \\ &= -\frac{\partial}{\partial z}\boldsymbol{E} + \nabla E_z \\ &= -\frac{\partial}{\partial z}\boldsymbol{E}_t + \nabla_t E_z \end{aligned} \tag{3.13a}$$

and

$$-j\omega\mu H_z - M_z = \boldsymbol{z}_0 \cdot (\nabla \times \boldsymbol{E}) = -\nabla_t \cdot (\boldsymbol{z}_0 \times \boldsymbol{E}). \tag{3.13b}$$

Similarly, for the second of (3.11), one has, by duality (see Chapter 2 Section III.4)

$$j\omega\varepsilon \boldsymbol{z}_0 \times \boldsymbol{E} + \boldsymbol{z}_0 \times \boldsymbol{J} = \nabla_t H_z - \frac{\partial}{\partial z}\boldsymbol{H}_t\,, \tag{3.14a}$$

$$j\omega\varepsilon E_z + J_z = \nabla_t \cdot (\boldsymbol{H} \times \boldsymbol{z}_0)\,. \tag{3.14b}$$

Upon replacing E_z in (3.13a), using (3.14b), one obtains

$$\begin{aligned}-\frac{\partial}{\partial z}\boldsymbol{E}_t &= j\omega\mu \boldsymbol{H}_t \times \boldsymbol{z}_0 - \frac{1}{j\omega\varepsilon}(\nabla_t\nabla_t \cdot \boldsymbol{H}_t \times \boldsymbol{z}_0 - \nabla_t J_z) + \boldsymbol{M}_t \times \boldsymbol{z}_0 \\ &= j\omega\mu\left(\mathscr{I} + \frac{\nabla_t\nabla_t}{k^2}\right) \cdot (\boldsymbol{H}_t \times \boldsymbol{z}_0) + \boldsymbol{M}_{te} \times \boldsymbol{z}_0\,,\end{aligned} \tag{3.15a}$$

and, by duality,

$$-\frac{\partial}{\partial z}\boldsymbol{H}_t = j\omega\varepsilon\left(\mathscr{I} + \frac{\nabla_t\nabla_t}{k^2}\right) \cdot (\boldsymbol{z}_0 \times \boldsymbol{E}_t) + \boldsymbol{z}_0 \times \boldsymbol{J}_{te}, \tag{3.15b}$$

where the equivalent transverse electric and magnetic current distributions are given, respectively, by

$$\boldsymbol{J}_{te} = \boldsymbol{J}_t - \boldsymbol{z}_0 \times \frac{\nabla_t M_z}{j\omega\mu} = \boldsymbol{J}_t + \frac{\nabla_t \times \boldsymbol{M}_z}{j\omega\mu}\,, \tag{3.16a}$$

$$\boldsymbol{M}_{te} = \boldsymbol{M}_t + \boldsymbol{z}_0 \times \frac{\nabla_t J_z}{j\omega\varepsilon} = \boldsymbol{M}_t - \frac{\nabla_t \times \boldsymbol{J}_z}{j\omega\varepsilon}\,. \tag{3.16b}$$

The transverse field equations (3.15) and (3.16), which admit z-dependent ε and μ, provide the basis for the treatment of field problems in uniform waveguides. They are completely descriptive of the total field equations (3.11), since from (3.13b) and (3.14b), the longitudinal components are derivable from the transverse components as (cf. (3.9a) and (3.9b))

$$j\omega\varepsilon E_z = \nabla_t \cdot (\boldsymbol{H}_t \times \boldsymbol{z}_0) - J_z\,, \tag{3.17a}$$

$$j\omega\mu H_z = \nabla_t \cdot (\boldsymbol{z}_0 \times \boldsymbol{E}_t) - M_z\,. \tag{3.17b}$$

The boundary condition (3.12), requiring the vanishing of the *total* tangential electric field on the perfectly conducting guide walls, can be restated in terms of the *transverse* field components as

$$\boldsymbol{\nu} \times \boldsymbol{E}_t = 0 \qquad \text{on } s, \tag{3.18a}$$

$$\nabla_t \cdot (\boldsymbol{H}_t \times \boldsymbol{z}_0) = 0 \qquad \text{on } s, \tag{3.18b}$$

where the second relation follows from (3.17a) upon assuming that $J_z = 0$ on s. This restriction, which requires the vanishing *on the boundary* of the z component of the *applied* electric current source, is of no practical consequence since an applied tangential electric current source on a perfectly conducting surface is "short-circuited" and cannot radiate a finite field.

III TE and TM Potentials

When it is possible to identify a preferred waveguiding direction, e.g. the longitudinal direction z, the field expressions (2.88b) simplify and the field can be separated into two parts, TE and TM.

E Type (TM) Potentials

A z-directed potential $\boldsymbol{\Pi}_e$

$$\boldsymbol{\Pi}_e = \boldsymbol{z}_0 \Pi_e(\boldsymbol{r}), \tag{3.19}$$

where $\boldsymbol{z}_0$ is a unit vector directed along z, generates a magnetic field contained entirely in the transverse plane and is therefore *Transverse Magnetic* TM or E type ($E_z \neq 0; H_z = 0$). It follows that for TM fields

$$\boldsymbol{\Pi}_h = 0 \tag{3.20}$$

and Π_e is a *scalar* potential, leading to substantial simplification in (2.88b). Decomposing

$$\nabla = \nabla_t + \boldsymbol{z}_0 \partial_z \tag{3.21}$$

and substituting (3.19) and (3.21) into (2.88b), we get

$$\boldsymbol{E} = \boldsymbol{E}_t + \boldsymbol{z}_0 E_z = \nabla_t \left(\partial_z \Pi_e\right) + \boldsymbol{z}_0 \left(\partial_z^2 \Pi_e + k^2 \Pi_e\right), \tag{3.22a}$$

$$\boldsymbol{H} = \boldsymbol{H}_t = -j\omega\varepsilon \boldsymbol{z}_0 \times \nabla_t \Pi_e . \tag{3.22b}$$

The scalar potential Π_e must satisfy the scalar Helmholtz equation in a source–free, locally or piecewise homogeneous region

$$\nabla^2 \Pi_e(\boldsymbol{r}) + k^2 \Pi_e(\boldsymbol{r}) = 0 \tag{3.23}$$

which, taking into account the transverse/longitudinal separability, i.e. $\Pi_e(\boldsymbol{r}) = \phi(\boldsymbol{\rho})\zeta(z)$, becomes the pair of scalar equations

$$\nabla_t^2 \phi(\boldsymbol{\rho}; k_t) + k_t^2 \phi(\boldsymbol{\rho}; k_t) = 0, \tag{3.24a}$$

$$\frac{d^2\zeta(z)}{dz^2} + \kappa^2 \zeta(z) = 0 \tag{3.24b}$$

linked by the wavenumber conservation condition (dispersion relation)

$$k_t^2 + \kappa^2 = k^2 , \tag{3.25}$$

where κ is the longitudinal wavenumber (propagation coefficient). In (3.24a), ϕ denotes a transverse eigenfunction, depending on the transverse wavenumbers (eigenvalue) k_t. When, as in closed metallic waveguides, the waveguide transverse cross–section is bounded, k_t can only take discrete values k_{ti},

$$k_{ti}^2 + \kappa_i^2 = k^2. \tag{3.26}$$

The corresponding transverse eigenfunction is denoted as $\phi_i(\boldsymbol{\rho})$, with i being the modal index. On the other hand, if the waveguide cross–section extends to infinity, k_t is a continuous variable indicative of a continuous spectrum as in the generic notation $\phi(\boldsymbol{\rho}; k_t)$.
When considering bounded cross-section waveguides with a discrete spectrum, by inserting $\Pi_e(\boldsymbol{r}) = \phi_i(\boldsymbol{\rho})\zeta_i(z)$ into (3.22) and making use of (3.24b,3.26) we have

$$E_{zi} = \kappa_i^2 \zeta_i(z)\phi_i(\boldsymbol{\rho}) , \tag{3.27a}$$

$$\boldsymbol{E}_{ti} = \nabla_t \phi_i(\boldsymbol{\rho}) \frac{d\zeta_i(z)}{dz} , \tag{3.27b}$$

$$\boldsymbol{H}_{ti} = -j\omega\varepsilon\zeta_i(z)\boldsymbol{z}_0 \times \nabla_t \phi_i(\boldsymbol{\rho}) . \tag{3.27c}$$

We obtain the appropriate *boundary condition* for $\phi_i(\boldsymbol{\rho})$ by observing from (3.27a) its proportionality to E_z, so that $E_z = 0$ on s in Figure 3.1 implies

$$\phi_i(\boldsymbol{\rho}) = 0 \quad on \ s. \tag{3.28}$$

It is convenient to normalize the transverse mode fields in (3.27) by introducing the orthonormal transverse vector eigenfunctions (primed quantities denote TM modes)

$$\mathbf{e}_i{}'(\boldsymbol{\rho}) = -\frac{\nabla_t \phi_i(\boldsymbol{\rho})}{k'_{ti}} , \tag{3.29a}$$

$$\boldsymbol{h}_i{}'(\boldsymbol{\rho}) = -\boldsymbol{z}_0 \times \mathbf{e}_i{}'(\boldsymbol{\rho}) \tag{3.29b}$$

with the corresponding modal vector fields in (3.27b,c) given by

$$\boldsymbol{E}'_{ti} = -k'_{ti} \frac{d\zeta_i(z)}{dz} \mathbf{e}_i{}'(\boldsymbol{\rho}) , \tag{3.30a}$$

$$\boldsymbol{H}'_{ti} = j\omega\varepsilon k'_{ti} \zeta_i(z) \boldsymbol{h}_i{}'(\boldsymbol{\rho}) . \tag{3.30b}$$

By introducing modal voltages $V_i'(z)$ and currents $I_i'(z)$

$$V_i'(z) = -k'_{ti}\frac{d\zeta_i(z)}{dz}\,, \tag{3.31a}$$

$$I_i'(z) = j\omega\varepsilon k'_{ti}\zeta_i(z) \tag{3.31b}$$

we may write

$$\boldsymbol{E}'_{ti} = V_i'(z)\boldsymbol{e}_i{}'(\boldsymbol{\rho})\,, \tag{3.32a}$$

$$\boldsymbol{H}'_{ti} = I_i'(z)\boldsymbol{h}_i{}'(\boldsymbol{\rho}). \tag{3.32b}$$

By taking the z-derivative of (3.31) we obtain the usual transmission-line equations

$$\frac{dI_i'(z)}{dz} = -j\kappa_i Y_i'(k'_{ti})V_i'(z)\,, \tag{3.33a}$$

$$\frac{dV_i'(z)}{dz} = -j\kappa_i Z_i'(k'_{ti})I_i'(z) \tag{3.33b}$$

with the modal impedance Z_i' and modal admittance Y_i' defined by

$$Z_i'(k_{ti}) = \frac{1}{Y_i'(k'_{ti})} = \frac{\kappa_i'}{\omega\varepsilon} \tag{3.34}$$

for TM modes.

H type (TE) Potentials

A potential $\boldsymbol{\Pi}_h$ directed along the z–axis

$$\boldsymbol{\Pi}_h = \boldsymbol{z}_0\Pi_h(\boldsymbol{r}), \tag{3.35}$$

generates an electric field contained entirely in the transverse plane, and is therefore *Transverse Electric* TE or H type ($H_z \neq 0, E_z = 0$). Thus, for TE modes

$$\boldsymbol{\Pi}_e = 0 \tag{3.36}$$

and Π_h is a *scalar* potential. We proceed in a manner dual to that for TM modes; for TE modes we label the corresponding quantities by a double prime.
The transverse vector eigenfunctions dual to those in (3.29) (i.e. $\boldsymbol{e}_i' \to \boldsymbol{h}_i''$, $\boldsymbol{h}_i' \to \boldsymbol{e}_i''$, $\phi_i \to \psi_i$) are given by

$$\boldsymbol{e}_i''(\boldsymbol{\rho}) = -\frac{\nabla_t\psi_i(\boldsymbol{\rho})}{k''_{ti}} \times \boldsymbol{z}_0, \tag{3.37a}$$

$$\boldsymbol{h}_i''(\boldsymbol{\rho}) = -\frac{\nabla_t\psi_i(\boldsymbol{\rho})}{k''_{ti}} \tag{3.37b}$$

with ψ_i satisfying the scalar eigenfunction equation

$$\nabla_t^2 \psi_i(\boldsymbol{\rho}) + k_{ti}''^2 \psi_i(\boldsymbol{\rho}) = 0 \tag{3.38}$$

subject to the PEC boundary condition $\boldsymbol{h}_i'' \cdot \boldsymbol{\nu} = 0$ on s (cf. (3.18)), which becomes (since $\nabla_t \cdot \boldsymbol{\nu} = \frac{\partial}{\partial \nu}$)

$$\frac{\partial \psi_i}{\partial \nu} = 0 \;\; on \;\; s. \tag{3.39}$$

Appealing again to duality, (3.32) become

$$\boldsymbol{E}_{ti}'' = V_i''(z)\boldsymbol{e}_i''(\boldsymbol{\rho}), \tag{3.40a}$$

$$\boldsymbol{H}_{ti}'' = I_i''(z)\boldsymbol{h}_i''(\boldsymbol{\rho}) \tag{3.40b}$$

with $V_i''(z)$ and $I_i''(z)$ satisfying the transmission-line equations

$$\frac{dV_i''(z)}{dz} = -j\kappa_i'' Z_i''(k_{ti}'') I_i''(z)\,, \tag{3.41a}$$

$$\frac{dI_i''(z)}{dz} = -j\kappa_i'' Y_i''(k_{ti}'') V_i''(z) \tag{3.41b}$$

with immittances now defined for the TE case as

$$Z_i''(k_{ti}'') = \frac{1}{Y_i''(k_{ti}'')} = \frac{\omega\mu}{\kappa_i''}\,. \tag{3.42}$$

The results obtained in this section are organized systematically into the architecture for modal representations of electromagnetic source–excited fields in Section IV below.

IV Modal Representations of the Fields and Their Sources

The vector electromagnetic field equations can be transformed into ordinary scalar differential equations on representation of the fields in terms of a complete orthonormal set of "guided" eigenfunctions. Single and double primes throughout denote H type TE and E type TM modes, respectively. For a perfectly conducting waveguide filled with a homogeneous, isotropic medium, a possible complete eigenvector set comprises both E TM mode functions $\boldsymbol{e}'(\boldsymbol{\rho})$, $\boldsymbol{h}'(\boldsymbol{\rho})$ and H TE mode functions $\boldsymbol{e}''(\boldsymbol{\rho})$, $\boldsymbol{h}''(\boldsymbol{\rho})$. In terms of the indicated mode functions, a representation of the independent transverse fields is given as

$$\boldsymbol{E}_t(\boldsymbol{r}) = \sum_i V_i'(z)\boldsymbol{e}_i'(\boldsymbol{\rho}) + \sum_i V_i''(z)\boldsymbol{e}_i''(\boldsymbol{\rho}), \tag{3.43a}$$

$$\boldsymbol{H}_t(\boldsymbol{r}) = \sum_i I_i'(z)\boldsymbol{h}_i'(\boldsymbol{\rho}) + \sum_i I_i''(z)\boldsymbol{h}_i''(\boldsymbol{\rho}), \tag{3.43b}$$

$$\boldsymbol{J}_{te}(\boldsymbol{r}) = \sum_i i_i'(z)\boldsymbol{e}_i'(\boldsymbol{\rho}) + \sum_i i_i''(z)\boldsymbol{e}_i''(\boldsymbol{\rho}), \tag{3.43c}$$

$$\boldsymbol{M}_{te}(\boldsymbol{r}) = \sum_i v_i'(z)\boldsymbol{h}_i'(\boldsymbol{\rho}) + \sum_i v_i''(z)\boldsymbol{h}_i''(\boldsymbol{\rho}), \tag{3.43d}$$

where i is in general a double index, and

$$\boldsymbol{h}_i = \boldsymbol{z}_0 \times \boldsymbol{e}_i. \tag{3.43e}$$

The specific form of the transverse vector eigenfunctions $\boldsymbol{e}_i$ and $\boldsymbol{h}_i$ is dependent on the shape of the guide cross-section and is, in general, defined by the following z-independent equations

$$\begin{aligned} \nabla_t\nabla_t \cdot \boldsymbol{e}_i' &= -k_{ti}'^2 \boldsymbol{e}_i' \qquad & \nabla_t\nabla_t \cdot \boldsymbol{h}_i'' &= -k_{ti}''^2 \boldsymbol{h}_i'', \\ \nabla_t\nabla_t \cdot \boldsymbol{h}_i' &= 0, & \nabla_t\nabla_t \cdot \boldsymbol{e}_i'' &= 0, \end{aligned} \tag{3.44}$$

subject, in accord with (3.18), to the boundary conditions on the curve s with normal $\boldsymbol{\nu}$ bounding the transverse cross section:

$$\boldsymbol{\nu} \times \boldsymbol{e}_i' = 0 = \nabla_t \cdot (\boldsymbol{h}_i' \times \boldsymbol{z}_0), \quad \boldsymbol{\nu} \times \boldsymbol{e}_i'' = 0 = \nabla_t \cdot (\boldsymbol{h}_i'' \times \boldsymbol{z}_0) \quad \text{on } s. \tag{3.45}$$

In view of (3.17) and (3.43), one obtains the longitudinal-field representations. One notes from (3.44) that only E modes contribute to the representation of E_z, while only H modes contribute to the representation of H_z,

$$j\omega\varepsilon E_z(\boldsymbol{r}) + J_z(\boldsymbol{r}) = \sum_i I_i'(z) \nabla_t \cdot \boldsymbol{e}_i'(\boldsymbol{\rho}), \tag{3.46a}$$

$$j\omega\mu H_z(\boldsymbol{r}) + M_z(\boldsymbol{r}) = \sum_i V_i''(z) \nabla_t \cdot \boldsymbol{h}_i''(\boldsymbol{\rho}). \tag{3.46b}$$

One notes from (3.45) that the vector mode functions in (3.43a) and (3.43b) individually satisfy the appropriate boundary conditions (3.18) on the transverse electromagnetic fields. Moreover, since *applied* electric and magnetic currents have no tangential or normal components, respectively, at a perfectly conducting surface, the representations for the source currents in (3.43c) and (3.43d) are likewise meaningful for realizable source current distributions on the boundary.
Upon applying the following transverse form of Green's theorem,[1]

$$\begin{aligned} &\iint_S dS[\boldsymbol{U} \cdot \nabla_t\nabla_t \cdot \boldsymbol{V} - \boldsymbol{V} \cdot \nabla_t\nabla_t \cdot \boldsymbol{U}] \\ &= \oint_s ds[(\boldsymbol{U} \cdot \boldsymbol{\nu})(\nabla_t \cdot \boldsymbol{V}) - (\boldsymbol{V} \cdot \boldsymbol{\nu})(\nabla_t \cdot \boldsymbol{U})] \ , \end{aligned} \tag{3.47a}$$

[1] Equation (3.47a) is obtained by applying the divergence theorem in the transverse cross section to the expression

$$\nabla_t \cdot [\boldsymbol{U}\nabla_t \cdot \boldsymbol{V} - \boldsymbol{V}\nabla_t \cdot \boldsymbol{U}] = \boldsymbol{U} \cdot \nabla_t\nabla_t \cdot \boldsymbol{V} - \boldsymbol{V} \cdot \nabla_t\nabla_t \cdot \boldsymbol{U}$$

where $\boldsymbol{U}$ and $\boldsymbol{V}$ are suitably continuous transverse vector functions, to the vector mode functions defined in (3.44), one deduces the orthogonality conditions over the cross-sectional domain S (normalization to unity is assumed):

$$\iint_S \boldsymbol{e}_i' \cdot \boldsymbol{e}_j'^* dS = \delta_{ij} = \iint_S \boldsymbol{e}_i'' \cdot \boldsymbol{e}_j''^* dS, \qquad \iint_S \boldsymbol{e}_i' \cdot \boldsymbol{e}_i''^* dS = 0, \tag{3.47b}$$

and similarly for the $\boldsymbol{h}_i$ functions. The asterisk denotes the complex conjugate,[2] and the Kronecker delta is defined as follows: $\delta_{ij} = 0, i \neq j; \delta_{ii} = 1$. In view of these orthonormality properties, the mode amplitudes in (3.43) are determined as follows:

$$V_i(z) = \iint_S \boldsymbol{E}_t(\boldsymbol{r}) \cdot \boldsymbol{e}_i^*(\boldsymbol{\rho})\, dS\,, \qquad I_i(z) = \iint_S \boldsymbol{H}_t(\boldsymbol{r}) \cdot \boldsymbol{h}_i^*(\boldsymbol{\rho})\, dS\,, \tag{3.48a}$$

$$v_i(z) = \iint_S \boldsymbol{M}_{te}(\boldsymbol{r}) \cdot \boldsymbol{h}_i^*(\boldsymbol{\rho})\, dS\,, \qquad i_i(z) = \iint_S \boldsymbol{J}_{te}(\boldsymbol{r}) \cdot \boldsymbol{e}_i^*(\boldsymbol{\rho})\, dS\,, \tag{3.48b}$$

where the distinguishing $'$ and $''$ have been omitted since the equations apply to both mode types. Utilizing the equivalent current definitions in (3.16) and employing the vector integration-by-parts formula (divergence theorem in two dimensions)

$$\iint_S dS\ \nabla_t f \cdot \boldsymbol{U} = -\iint_S dS\ f \nabla_t \cdot \boldsymbol{U} + \oint_s ds f(\boldsymbol{U} \cdot \boldsymbol{\nu}) \tag{3.49}$$

with f and $\boldsymbol{U}$ suitably continuous scalar and vector functions, one may reexpress the integrals of (3.48). The contribution to the gradient integrals from the bounding contour s vanishes in view of the boundary condition $\boldsymbol{h}_i \cdot \boldsymbol{\nu} = 0$ [(3.45)] and the specification $J_z = 0$ on s, so (3.48b) become

$$v_i(z) = \iint_S \boldsymbol{M}(\boldsymbol{r}) \cdot \boldsymbol{h}_i^*(\boldsymbol{\rho}) dS + Z_i^* \iint_S \boldsymbol{J}(\boldsymbol{r}) \cdot \boldsymbol{e}_{zi}^*(\boldsymbol{\rho}) dS, \tag{3.50a}$$

$$i_i(z) = \iint_S \boldsymbol{J}(\boldsymbol{r}) \cdot \boldsymbol{e}_i^*(\boldsymbol{\rho}) dS + Y_i^* \iint_S \boldsymbol{M}(\boldsymbol{r}) \cdot \boldsymbol{h}_{zi}^*(\boldsymbol{\rho}) dS, \tag{3.50b}$$

where

$$Y_i'' \boldsymbol{h}_{zi}''(\boldsymbol{\rho}) \equiv \boldsymbol{z}_0 \frac{\nabla_t \cdot \boldsymbol{h}_i''(\boldsymbol{\rho})}{j\omega\mu}, \qquad \boldsymbol{h}_{zi}' \equiv 0, \tag{3.50c}$$

$$Z_i' \boldsymbol{e}_{zi}'(\boldsymbol{\rho}) \equiv \boldsymbol{z}_0 \frac{\nabla_t \cdot \boldsymbol{e}_i'(\boldsymbol{\rho})}{j\omega\varepsilon}, \qquad \boldsymbol{e}_{zi}'' \equiv 0. \tag{3.50d}$$

---

[2] Although $k_{ti}'^2$ and $k_{ti}''^2$ are real (which guarantees real eigenfunctions), it may be convenient to employ a complex decomposition (e.g. $cos(\alpha x) = 1/2[\exp(j\alpha x) + \exp(-j\alpha x)]$. Therefore, the orthogonality condition involves the complex conjugate function.

The vanishing of $\boldsymbol{h}'_{zi}$ (for E modes) and of $\boldsymbol{e}''_{zi}$ (for H modes) follows directly from (3.44). The introduction of the characteristic impedance and admittance Z'_i and Y''_i [defined explicitly in (3.51d)] serves to highlight in a physical sense the contributions of the various integrals as either voltages or currents. It is to be noted that the formulations in (3.50) do not require differentiability of J_z and M_z in the cross section S as implied in (3.16) and (3.48b). By inserting the modal representations (3.43) into the transverse field equations (3.15), interchanging the order of summation and differentiation, making use of (3.44), and equating like coefficients of the mode functions $\boldsymbol{e}_i$, and $\boldsymbol{h}_i$, one obtains the desired transmission-line (TL) equations for the E and H mode amplitudes as

$$-\frac{dV_i}{dz} = j\kappa_i Z_i I_i + v_i, \tag{3.51a}$$

$$-\frac{dI_i}{dz} = j\kappa_i Y_i V_i + i_i, \tag{3.51b}$$

where the modal characteristic impedance Z_i (admittance Y_i) and the modal propagation constant κ_i, are defined as follows:

E modes:

$$Z'_i = \frac{1}{Y'_i} = \frac{\kappa'_i}{\omega\varepsilon}, \quad \kappa'_i = \sqrt{k^2 - k'^2_{ti}} = -j\sqrt{k'^2_{ti} - k^2}, \tag{3.51c}$$

H modes:

$$Z''_i = \frac{1}{Y''_i} = \frac{\omega\mu}{\kappa''_i}, \quad \kappa''_i = \sqrt{k^2 - k''^2_{ti}} = -j\sqrt{k''^2_{ti} - k^2}. \tag{3.51d}$$

Here, $k^2 = \omega^2\mu\varepsilon$, and both μ and ε may be functions of z. The form of (3.51a) and (3.51b) permits identification of V_i and I_i as transmission-line voltages and currents, respectively. The choice of sign on the square roots in (3.51) assures the damping of non-propagating modes (κ_i imaginary) away from the source region for the assumed time dependence $\exp(+j\omega t)$. The evaluation of the source voltage v_i and current i_i amplitudes follows directly from the *specified* electric and magnetic source currents $\boldsymbol{J}$ and $\boldsymbol{M}$ via (3.50a) and (3.50b). Solutions of the network–oriented TL equations (3.51a) and (3.51b) for various stratifications and terminations in the z domain are discussed next.

V Scalarization and Modal Representation of Dyadic Green's Functions in Uniform Regions

Solutions for the vector electromagnetic field excited by prescribed sources in a uniform waveguide region bounded by perfectly conducting walls (if any) and filled with a transversely homogeneous material follow from the representations in (3.43) and (3.46); the vector mode functions are evaluated from (3.44) and the modal amplitudes from (3.51), subject to appropriate boundary conditions in the z domain. Solution of the vector eigenvalue problems in (3.44) is facilitated by introduction of scalar mode functions. The scalarization achieved in this manner may be utilized to define E and H mode (Hertz) potentials from which the electromagnetic fields themselves can be derived. For point-source excitation, these potentials are equivalent to scalar Green's functions. The procedure discussed below yields explicit expressions for these functions and thereby *solves* the scalar potential problems. We first express vector mode functions in terms of scalar mode functions, and then scalarize the overall field representation.

V.1 Mode Functions

In representing the transverse electric vector field $\boldsymbol{E}_t$ in (3.43a) in terms of two independent vector mode sets $\{\boldsymbol{e}'_i\}$ and $\{\boldsymbol{e}''_i\}$, use has been made of a theorem which states that any *transverse* vector can be decomposed into two parts, one of which is with zero divergence (solenoidal) and the other of which is with zero curl (irrotational). The vector set $\{\boldsymbol{e}'_i\}$ is irrotational (i.e., $\nabla_t \times \boldsymbol{e}'_i = 0$ in S), while the vector set $\{\boldsymbol{e}''_i\}$ is solenoidal (i.e., $\nabla_t \cdot \boldsymbol{e}''_i = 0$ in S) [see also (3.44)]. In view of these properties, the vector mode functions $\boldsymbol{e}'_i$ and $\boldsymbol{e}''_i$ can be represented as gradients and curls of scalar functions ϕ_i and ψ_i as follows (recall that curl-grad and div-curl $\equiv 0$)

$$\boldsymbol{e}'_i(\boldsymbol{\rho}) = -\frac{\nabla_t \phi_i(\boldsymbol{\rho})}{k'_{ti}}, \tag{3.52a}$$

$$\boldsymbol{e}''_i(\boldsymbol{\rho}) = -\frac{\nabla_t \psi_i(\boldsymbol{\rho})}{k''_{ti}} \times \boldsymbol{z}_0, \tag{3.52b}$$

and, consequently,

$$\boldsymbol{h}'_i(\boldsymbol{\rho}) = -\boldsymbol{z}_0 \times \frac{\nabla_t \phi_i(\boldsymbol{\rho})}{k'_{ti}}, \tag{3.52c}$$

$$\boldsymbol{h}''_i(\boldsymbol{\rho}) = -\frac{\nabla_t \psi_i(\boldsymbol{\rho})}{k''_{ti}}. \tag{3.52d}$$

By (3.52) and (3.44), the mode functions ϕ_i, and ψ_i are defined by the two scalar eigenvalue problems (note that $\nabla_t^2 = \nabla_t \cdot \nabla_t$)

$$\nabla_t^2 \phi_i + {k'_{ti}}^2 \phi_i = 0 \qquad \text{in } S, \tag{3.53a}$$

$$\begin{aligned} \phi_i &= 0 \quad \text{on } s \text{ if } k'_{ti} \neq 0, \\ \frac{\partial \phi_i}{\partial s} &= 0 \quad \text{on } s \text{ if } k'_{ti} = 0 \qquad \text{(TEM mode)}, \end{aligned} \tag{3.53b}$$

and

$$\nabla_t^2 \psi_i + k_{ti}''^2 \psi_i = 0 \qquad \text{in } S, \tag{3.53c}$$

$$\frac{\partial \psi_i}{\partial \nu} = 0 \qquad \text{on } s. \tag{3.53d}$$

The vector mode functions for the TEM (transverse electromagnetic) case are determined via

$$\boldsymbol{e}_0'(\boldsymbol{\rho}) = \boldsymbol{h}_0'(\boldsymbol{\rho}) \times \boldsymbol{z}_0 = -\nabla_t \phi_0(\boldsymbol{\rho}), \tag{3.54}$$

where $\phi_0(\boldsymbol{\rho})$ is the solution of (3.53a) with $k_{ti}' = 0$, with the normalization

$$\iint_S e_0'^2(\boldsymbol{\rho})\, dS = 1\,. \tag{3.55}$$

VI Fields in Source-Free, Homogeneous Regions

Using (3.52) and assuming interchangeability of summation and differentiation operations, one may write (3.43a) and (3.43b) as

$$\boldsymbol{E}_t(\boldsymbol{r}) = -\nabla_t V'(\boldsymbol{r}) - \nabla_t V''(\boldsymbol{r}) \times \boldsymbol{z}_0, \tag{3.56a}$$

$$\boldsymbol{H}_t(\boldsymbol{r}) \times \boldsymbol{z}_0 = -\nabla_t I'(\boldsymbol{r}) - \nabla_t I''(\boldsymbol{r}) \times \boldsymbol{z}_0, \tag{3.56b}$$

where the potential functions $V'(\boldsymbol{r})$, $I'(\boldsymbol{r})$ and $V''(\boldsymbol{r})$, $I''(\boldsymbol{r})$ are defined as follows:

$$V'(\boldsymbol{r}) = \sum_i V_i'(z) \frac{\phi_i(\boldsymbol{\rho})}{k_{ti}'}, \qquad V''(\boldsymbol{r}) = \sum_i V_i''(z) \frac{\psi_i(\boldsymbol{\rho})}{k_{ti}''}, \tag{3.57a}$$

$$I'(\boldsymbol{r}) = \sum_i I_i'(z) \frac{\phi_i(\boldsymbol{\rho})}{k_{ti}'}, \qquad I''(\boldsymbol{r}) = \sum_i I_i''(z) \frac{\psi_i(\boldsymbol{\rho})}{k_{ti}''}. \tag{3.57b}$$

From (3.56) and (3.17), the electromagnetic fields can be expressed at any source-free point where ε and μ are *non-variable* as[3]

$$\boldsymbol{E}(\boldsymbol{r}) = \frac{1}{j\omega\varepsilon} \nabla \times \nabla \times [\boldsymbol{z}_0 V'(\boldsymbol{r})] - \nabla \times [\boldsymbol{z}_0 V''(\boldsymbol{r})], \tag{3.58a}$$

$$\boldsymbol{H}(\boldsymbol{r}) = \nabla \times [\boldsymbol{z}_0 V'(\boldsymbol{r})] + \frac{1}{j\omega\mu} \nabla \times \nabla \times [\boldsymbol{z}_0 V''(\boldsymbol{r})]. \tag{3.58b}$$

[3] It should be pointed out that the scalar eigenfunctions ϕ_i and ψ_i, like the vector eigenfunctions $\boldsymbol{e}_i'$ and $\boldsymbol{e}_i''$, each form an orthonormal set (see Section 3.2). Normalization of these scalar eigenfunctions differs from that used in reference [14]. The relation between the eigenfunctions here and those in reference [14] is the following:

$$k_{ti}' [\phi_i]_{\text{ref.1}} = \phi_i\,, \qquad k_{ti}'' [\psi_i]_{\text{ref.1}} = \psi_i\,.$$

The two independent functions $I'(\boldsymbol{r})$ and $V''(\boldsymbol{r})$ suffice to determine the total fields via (3.17). In a *source-free region*, $V'(\boldsymbol{r})$ and $I''(\boldsymbol{r})$ are obtainable from $I'(\boldsymbol{r})$ and $V''(\boldsymbol{r})$, respectively, by differentiation with respect to z, as is evident from the transmission-line equations (3.51). Thus,

$$V'(\boldsymbol{r}) = \sum_i \frac{1}{-j\kappa_i' Y_i'} \frac{dI_i'(z)}{dz} \frac{\phi_i(\boldsymbol{\rho})}{k_{ti}'} = \frac{1}{-j\omega\varepsilon} \frac{\partial}{\partial z} I'(\boldsymbol{r}), \tag{3.59a}$$

and, similarly,

$$I''(\boldsymbol{r}) = \frac{1}{-j\omega\mu} \frac{\partial}{\partial z} V''(\boldsymbol{r}). \tag{3.59b}$$

Equations (3.53) and (3.51) may be used to verify that in a source-free, homogeneous region, the potentials I' and V'' given by (3.57) satisfy the Helmholtz equations

$$(\nabla^2 + k^2) \begin{Bmatrix} I' \\ V'' \end{Bmatrix} = 0. \tag{3.60}$$

The potential functions (V', I') and (V'', I'') satisfying (3.60) are of the Hertz–potential type, as can be seen by comparison with $\boldsymbol{\Pi}_e$ in (3.19) and (3.23) and its dual $\boldsymbol{\Pi}_h$ in (3.35) etc., respectively.

VII Green's Functions for the Transmission-Line Equations

To obtain explicit solutions for the potentials in source regions, it is necessary to relate the modal coefficients in (3.57) to their excitations. Within this context, it is convenient to introduce modal Green's functions, which characterize the response at z due to a point source at z'. In view of the linearity of the TL equations (3.51), one can obtain the voltage and current solutions at any point z by superposing separate contributions from appropriately weighted point voltage and current generators distributed along points z'. Thus,

$$V(z) = -\int dz'\, T^V(z,z')v(z') - \int dz'\, Z(z,z')i(z'), \tag{3.61a}$$

$$I(z) = -\int dz'\, Y(z,z')v(z') - \int dz'\, T^I(z,z')i(z'), \tag{3.61b}$$

where the mode subscript i has been omitted. Equations (3.61) reduce the problem to that of determining $T^V(z,z')$, $Y(z,z')$ and $Z(z,z')$, $T^I(z,z')$, whose significance as modal Green's functions is evident: $-T^V(z,z')$ and $-Z(z,z')$ are the *voltage* responses at z due, respectively, to a unit voltage and current source (generator) at z', while $-Y(z,z')$ and $-T^I(z,z')$ are the corresponding *current* responses to the same excitations. Thus, if in (3.51), one sets $v(z) = -\delta(z - z')$ and $i(z) = 0$, there results

$$-\frac{d}{dz} T^V(z, z') = j\kappa Z Y(z, z') - \delta(z - z'), \tag{3.62a}$$

$$-\frac{d}{dz} Y(z, z') = j\kappa Y T^V(z, z'), \tag{3.62b}$$

and, if $v = 0$, $i = -\delta(z - z')$,

$$-\frac{d}{dz} Z(z, z') = j\kappa Z T^I(z, z'), \tag{3.62c}$$

$$-\frac{d}{dz} T^I(z, z') = j\kappa Y Z(z, z') - \delta(z - z'), \tag{3.62d}$$

subject to as-yet-unspecified boundary conditions at the z terminations.
The modal Green's functions defined in (3.62) satisfy reciprocity properties when κ and Z are either constant or z–dependent. Consider a given terminated transmission line to be excited by two separate source distributions: the first, $v(z)$, $i(z)$, giving rise to $V(z)$, $I(z)$; and the second, $\hat{v}(z)$, $\hat{i}(z)$, giving rise to $\hat{V}(z)$, $\hat{I}(z)$. Both sets satisfy the TL equations:

$$-\frac{dV}{dz} = j\kappa Z I + v, \tag{3.63a}$$

$$-\frac{dI}{dz} = j\kappa Y V + i, \tag{3.63b}$$

and

$$-\frac{d\hat{V}}{dz} = j\kappa Z \hat{I} + \hat{v} \tag{3.63c}$$

$$-\frac{d\hat{I}}{dz} = j\kappa Y \hat{V} + \hat{i}. \tag{3.63d}$$

Upon multiplying (3.63a)–(3.63d) by $\hat{I}$, $\hat{V}$, I, V, respectively, subtracting the sum of the resulting (3.63a) and (3.63d) from the sum of (3.63b) and (3.63c), and integrating over z between the limits z_1 and z_2, one obtains

$$(\hat{V} I - \hat{I} V)_{z_1}^{z_2} = \int_{z_1}^{z_2} dz (v \hat{I} + \hat{i} V - i \hat{V} - \hat{v} I). \tag{3.64}$$

subject to the same terminal conditions at z_1 and z_2

$$V(z_{1,2}) = \mp Z(z_{1,2}) I(z_{1,2}), \qquad \hat{V}(z_{1,2}) = \mp Z(z_{1,2}) \hat{I}(z_{1,2}), \tag{3.65}$$

where $Z(z_{1,2})$ are terminal impedances[4]. Thus, the left-hand side of (3.64), expressing the difference between the values at z_2 and z_1 of the bracketed quantity, vanishes and one obtains the reciprocity relation

$$\int_{z_1}^{z_2} dz(v\hat{I} + \hat{i}V - i\hat{V} - \hat{v}I) = 0. \tag{3.66}$$

To apply the reciprocity condition (3.66) to the modal Green's functions defined in (3.62), one selects the following special source distributions:

$$\begin{gathered}
v = \hat{v} = 0, \quad i = -\delta(z - z'), \quad \hat{i} = -\delta(z - z''); \\
V \to Z(z, z'), \quad \hat{V} \to Z(z, z''), \\
i = \hat{i} = 0, \quad v = -\delta(z - z'), \quad \hat{v} = -\delta(z - z''); \\
I \to Y(z, z'), \quad \hat{I} \to Y(z, z''), \\
v = \hat{i} = 0, \quad i = -\delta(z - z'), \quad \hat{v} = -\delta(z - z''); \\
I \to T^I(z, z'), \quad \hat{V} \to T^V(z, z''),
\end{gathered}$$

whence one obtains the following reciprocity theorems:

$$Z(z'', z') = Z(z', z''), \tag{3.68a}$$

$$Y(z'', z') = Y(z', z''), \tag{3.68b}$$

$$T^I(z'', z') = -T^V(z', z''). \tag{3.68c}$$

In view of the reciprocity relation (3.68c) between T^I and T^V, one deduces from (3.62) the important fact that the general solution for the voltage and current in a source-free region can be expressed *solely* in terms of *either* $Y(z, z')$ or $Z(z, z')$. Suppose we have found $Y(z, z')$; then T^V is obtained from (3.62b). Because of the reciprocity theorem, a knowledge of T^V implies the knowledge of T^I, which in turn determines $Z(z, z')$ via (3.62d), provided that $z \neq z'$ (i.e., away from the source). Thus, all the required information is contained in $Y(z, z')$; an alternative statement applies for $Z(z, z')$. Because of the fundamental role played by the current (i.e., the E_z field component) in the case of E modes, it is usually convenient to determine E mode solutions from $Y(z, z')$; by duality, the Green's function $Z(z, z')$ is usually more convenient for H mode quantities.

VIII Modal Representations of the Dyadic Green's Functions in a Piecewise Homogeneous Medium

The electromagnetic fields radiated by point current excitations are conveniently expressed in terms of dyadic Green's functions. In this section we derive modal

[4] Note that in this section, Z, $Z(z_\alpha)$, and $Z(z, z')$ denote, respectively, the modal characteristic impedance, the terminating impedance at z_α, and the voltage Green's function for the ith mode.

solutions for the dyadic Green's functions in regions whose properties are constant along the z direction and show how the dyadic Green's functions can be related to scalar Green's functions.

From (3.58) one notes that the electromagnetic fields $\boldsymbol{E}(\boldsymbol{r})$ and $\boldsymbol{H}(\boldsymbol{r})$ exterior to source regions can be expressed in terms of the scalar potential functions $I'(\boldsymbol{r})$ and $V''(\boldsymbol{r})$ defined in (3.57). If the assumed sources are electric and magnetic current elements situated at the point $\boldsymbol{r}'$,

$$\boldsymbol{J}(\boldsymbol{r}) = \boldsymbol{J}^0\delta(\boldsymbol{r}-\boldsymbol{r}'), \qquad \boldsymbol{M}(\boldsymbol{r}) = \boldsymbol{M}^0\delta(\boldsymbol{r}-\boldsymbol{r}'), \tag{3.69}$$

where $\boldsymbol{J}^0$ and $\boldsymbol{M}^0$ are arbitrarily oriented constant vectors, then the modal representations for I' and V'' in (3.57) can be simplified. Consider first the E mode TM current $I_i'(z)$ occurring in the representation for the E mode TM current potential $I'(\boldsymbol{r})$ in (3.57b). Upon recalling the definitions for the transmission-line Green's functions $Y_i(z,z')$ and $T_i^I(z,z')$ in (3.61b), one notes that for a point source

$$I_i'(z,z') = -Y_i'(z,z')v_i'(z') - T_i'^I(z,z')i_i'(z'), \tag{3.70}$$

where the dependence of $I_i'(z)$ on z' has been indicated explicitly and the subscripts have been inserted to highlight the modal character of the various quantities. It will be desirable to have $T_i'^I(z,z')$ expressed in terms of $Y_i'(z,z')$. From (3.68c), (3.62b), and (3.68b), one finds that

$$T_i^I(z,z') = -T_i^V(z',z) = \frac{1}{j\kappa_i Y_i}\frac{d}{dz'}Y_i(z',z) = \frac{1}{j\kappa_i Y_i}\frac{d}{dz'}Y_i(z,z'). \tag{3.71}$$

Since $\kappa_i'Y_i' = \omega\varepsilon$ for E modes [see (3.51c)], one obtains, instead of (3.70),

$$I_i'(z,z') = -\left[v_i'(z') + \frac{1}{j\omega\varepsilon}i_i'(z')\frac{d}{dz'}\right]Y_i'(z,z'). \tag{3.72}$$

In a similar manner, one can show that the H mode voltages $V_i''(z)$, occurring in the representation of the voltage potential function $V''(\boldsymbol{r})$ in (3.57a), can be expressed in a manner dual to that in (3.72):

$$V_i''(z,z') = -\left[i_i''(z') + \frac{1}{j\omega\mu}v_i''(z')\frac{d}{dz'}\right]Z_i''(z,z')\,. \tag{3.73}$$

Since $\delta(\boldsymbol{r}-\boldsymbol{r}') = \delta(\boldsymbol{\rho}-\boldsymbol{\rho}')\delta(z-z')$ in (3.69), the source terms v_i and i_i defined in terms of $\boldsymbol{J}$ and $\boldsymbol{M}$ by (3.50), take on the following simple form:

$$v_i(z) = v_i(z')\delta(z-z'), \qquad i_i(z) = i_i(z')\delta(z-z'), \tag{3.74a}$$

$$v_i(z') = \boldsymbol{h}_i^*(\boldsymbol{\rho}')\cdot\boldsymbol{M}^0 + Z_i^*\boldsymbol{e}_{zi}^*(\boldsymbol{\rho}')\cdot\boldsymbol{J}^0, \tag{3.74b}$$

$$i_i(z') = \boldsymbol{e}_i^*(\boldsymbol{\rho}')\cdot\boldsymbol{J}^0 + Y_i^*\boldsymbol{h}_{zi}^*(\boldsymbol{\rho}')\cdot\boldsymbol{M}^0. \tag{3.74c}$$

Upon substituting the scalar mode functions via (3.52), one finds that for E modes,

$$
\begin{aligned}
&-\left[v_i'(z') + \frac{1}{j\omega\varepsilon} i_i'(z') \frac{d}{dz'}\right] \\
&= \left[(\boldsymbol{z}_0 \times \nabla_t') \frac{\phi_i^*(\boldsymbol{\rho}')}{k_{ti}'}\right] \cdot \boldsymbol{M}^0 - \frac{1}{j\omega\varepsilon}\left[\left(\boldsymbol{z}_0 \nabla_t'^2 - \nabla_t' \frac{\partial}{\partial z'}\right) \frac{\phi_i^*(\boldsymbol{\rho}')}{k_{ti}'}\right] \cdot \boldsymbol{J}^0, \qquad (3.75)
\end{aligned}
$$

where ∇_t' denotes differentiation with respect to the primed coordinates $\boldsymbol{\rho}'$. In view of the vector identities

$$
\boldsymbol{z}_0 \times \nabla_t' \bar{\varphi} = -\nabla' \times (\boldsymbol{z}_0 \bar{\varphi}) \to -(\nabla' \times \boldsymbol{z}_0)\bar{\varphi} \qquad (3.76a)
$$

and

$$
\begin{aligned}
\left(\nabla_t' \frac{\partial}{\partial z'} - \boldsymbol{z}_0 \nabla_t'^2\right) \bar{\varphi} &= \left(\nabla' \frac{\partial}{\partial z'} - \boldsymbol{z}_0 \nabla'^2\right) \bar{\varphi} \\
&= \nabla'(\nabla' \cdot \boldsymbol{z}_0 \bar{\varphi}) - \nabla'^2(\boldsymbol{z}_0 \bar{\varphi}) \to (\nabla' \times \nabla' \times \boldsymbol{z}_0)\bar{\varphi},
\end{aligned} \qquad (3.76b)
$$

where $\bar{\varphi}$ is a scalar function of $\boldsymbol{\rho}'$, one obtains the following concise expression for $I'(\boldsymbol{r})$ after substituting (3.72)–(3.76) into (3.57b):

$$
I'(\boldsymbol{r}) = (\nabla' \times \nabla' \times \boldsymbol{z}_0)\mathscr{S}'(\boldsymbol{r}, \boldsymbol{r}') \cdot \boldsymbol{J}^0 - j\omega\varepsilon(\nabla' \times \boldsymbol{z}_0)\mathscr{S}'(\boldsymbol{r}, \boldsymbol{r}') \cdot \boldsymbol{M}^0, \qquad (3.77a)
$$

where

$$
j\omega\varepsilon\mathscr{S}'(\boldsymbol{r}, \boldsymbol{r}') = \sum_i \frac{\phi_i(\boldsymbol{\rho})\phi_i^*(\boldsymbol{\rho}')}{k_{ti}'^2} Y_i'(\boldsymbol{z}, \boldsymbol{z}'). \qquad (3.77b)
$$

The meaning of the operations $\nabla' \times \boldsymbol{z}_0$ and $\nabla' \times \nabla' \times \boldsymbol{z}_0$ is defined in (3.76a) and (3.76b), respectively. Equations (3.77) evidently are valid only when $k_{ti}' \neq 0$ (i.e., any possible TEM modes are excluded).[5] If the waveguide structure can support one or more TEM modes, the contribution to the radiated fields from these modes must be taken into account separately [see footnote to (3.53b)]. For the H mode potential function $V''(\boldsymbol{r})$ in (3.57a) one obtains by analogous considerations the dual representation

$$
V''(\boldsymbol{r}) = j\omega\mu(\nabla' \times \boldsymbol{z}_0)\mathscr{S}''(\boldsymbol{r}, \boldsymbol{r}') \cdot \boldsymbol{J}^0 + (\nabla' \times \nabla' \times \boldsymbol{z}_0)\mathscr{S}''(\boldsymbol{r}, \boldsymbol{r}') \cdot \boldsymbol{M}^0, \qquad (3.78a)
$$

where

[5] The interchange of operations of summation and differentiation, assumed valid in deriving (3.77) from (3.57), may not be permissible in certain problems involving continuous spectra or eigenfunctions. [Similar remarks apply to (3.78).] In these instances, the above expressions are to be considered as formal and must be properly interpreted [see the last paragraph in this section for related comments pertaining to the operator $1/\nabla_t^2$].

$$j\omega\mu\mathscr{S}''(\boldsymbol{r},\boldsymbol{r}') = \sum_i \frac{\psi_i(\boldsymbol{\rho})\psi_i^*(\boldsymbol{\rho}')}{k_{ti}''^2} Z_i''(\boldsymbol{z},\boldsymbol{z}'), \tag{3.78b}$$

and ψ_i are the scalar H mode functions defined in (3.53).
Upon substituting the representations for $I'(\boldsymbol{r})$ and $V''(\boldsymbol{r})$ from (3.77) and (3.78) into (3.58), one obtains the desired formulation for the electromagnetic fields observed at $\boldsymbol{r}$ due to vector point-source excitations of electric and magnetic currents at $\boldsymbol{r}'$ as in (3.69):

$$\boldsymbol{E}(\boldsymbol{r},\boldsymbol{r}') = -\mathscr{Z}(\boldsymbol{r},\boldsymbol{r}')\cdot\boldsymbol{J}^0 - \mathscr{T}_e(\boldsymbol{r},\boldsymbol{r}')\cdot\boldsymbol{M}^0, \tag{3.79a}$$

$$\boldsymbol{H}(\boldsymbol{r},\boldsymbol{r}') = -\mathscr{T}_m(\boldsymbol{r},\boldsymbol{r}')\cdot\boldsymbol{J}^0 - \mathscr{Y}(\boldsymbol{r},\boldsymbol{r}')\cdot\boldsymbol{M}^0, \tag{3.79b}$$

where $\mathscr{Z}$, $\mathscr{Y}$ and $\mathscr{T}_e$, $\mathscr{T}_m$ are the dyadic impedance, admittance, and electric and magnetic transfer functions, respectively [with $\boldsymbol{r} \neq \boldsymbol{r}'$]:

$$\begin{aligned} -j\omega\varepsilon\mathscr{Z}(\boldsymbol{r},\boldsymbol{r}') &= (\nabla\times\nabla\times\boldsymbol{z}_0)(\nabla'\times\nabla'\times\boldsymbol{z}_0)\mathscr{S}'(\boldsymbol{r},\boldsymbol{r}') \\ &\quad + k^2(\nabla\times\boldsymbol{z}_0)(\nabla'\times\boldsymbol{z}_0)\mathscr{S}''(\boldsymbol{r},\boldsymbol{r}'), \end{aligned} \tag{3.80a}$$

$$\begin{aligned} -j\omega\mu\mathscr{Y}(\boldsymbol{r},\boldsymbol{r}') &= (\nabla\times\nabla\times\boldsymbol{z}_0)(\nabla'\times\nabla'\times\boldsymbol{z}_0)\mathscr{S}''(\boldsymbol{r},\boldsymbol{r}') \\ &\quad + k^2(\nabla\times\boldsymbol{z}_0)(\nabla'\times\boldsymbol{z}_0)\mathscr{S}'(\boldsymbol{r},\boldsymbol{r}'), \end{aligned} \tag{3.80b}$$

$$\begin{aligned} \mathscr{T}_e(\boldsymbol{r},\boldsymbol{r}') &= (\nabla\times\nabla\times\boldsymbol{z}_0)(\nabla'\times\boldsymbol{z}_0)\mathscr{S}'(\boldsymbol{r},\boldsymbol{r}') \\ &\quad + (\nabla\times\boldsymbol{z}_0)(\nabla'\times\nabla'\times\boldsymbol{z}_0)\mathscr{S}''(\boldsymbol{r},\boldsymbol{r}'), \end{aligned} \tag{3.80c}$$

$$\begin{aligned} -\mathscr{T}_m(\boldsymbol{r},\boldsymbol{r}') &= (\nabla\times\nabla\times\boldsymbol{z}_0)(\nabla'\times\boldsymbol{z}_0)\mathscr{S}''(\boldsymbol{r},\boldsymbol{r}') \\ &\quad + (\nabla\times\boldsymbol{z}_0)(\nabla'\times\nabla'\times\boldsymbol{z}_0)\mathscr{S}'(\boldsymbol{r},\boldsymbol{r}'), \end{aligned} \tag{3.80d}$$

where $k^2 = \omega^2\mu\varepsilon =$ constant. Via (3.80), the dyadic Green's functions are expressed in terms of scalar functions $\mathscr{S}'$ and $\mathscr{S}''$ in what appears to be a fundamental form. The symmetry inherent in the expressions is to be noted. In (3.84b) and (3.85b) the functions $-\nabla_t'^2\mathscr{S}'$ and $-\nabla_t'^2\mathscr{S}''$ are shown to be scalar Green's functions that satisfy (3.88) and (3.89). Since from (3.68), $Y_i'(z,z') = Y_i'(z',z)$ and $Z_i''(z,z') = Z_i''(z',z)$, it follows from the modal representations for $\mathscr{S}'$ and $\mathscr{S}''$ in (3.77b) and (3.78b), respectively, that for *real* ϕ_i and ψ_i[6]

$$\mathscr{S}'(\boldsymbol{r},\boldsymbol{r}') = \mathscr{S}'(\boldsymbol{r}',\boldsymbol{r}), \qquad \mathscr{S}''(\boldsymbol{r},\boldsymbol{r}') = \mathscr{S}''(\boldsymbol{r}',\boldsymbol{r}), \tag{3.81}$$

whence, from (3.80),

$$\mathscr{Z}(\boldsymbol{r},\boldsymbol{r}') = \widetilde{\mathscr{Z}}(\boldsymbol{r}',\boldsymbol{r}), \quad \mathscr{Y}(\boldsymbol{r},\boldsymbol{r}') = \widetilde{\mathscr{Y}}(\boldsymbol{r}',\boldsymbol{r}), \quad \mathscr{T}_e(\boldsymbol{r},\boldsymbol{r}') = -\widetilde{\mathscr{T}}_m(\boldsymbol{r}',\boldsymbol{r}), \tag{3.82}$$

---

[6] Although not always convenient, the mode functions ϕ_i and ψ_i in regions bounded either by perfectly conducting walls, or else unbounded, can always be chosen real. Only such regions, wherein k_{ti}^2 is real, are considered above.

where the tilde ($\tilde{\ }$) denotes the transposed dyadics. These relations represent reciprocity conditions valid for $\boldsymbol{r} \neq \boldsymbol{r}'$. (To include also the point $\boldsymbol{r} = \boldsymbol{r}'$, (3.80) must be modified as in (1.1.38) or (1.1.49) of [14]).

Equations (3.77) and (3.78) simplify considerably for the case of longitudinal sources,

$$\boldsymbol{J}^0 = \boldsymbol{z}_0 J^0, \qquad \boldsymbol{M}^0 = \boldsymbol{z}_0 M^0. \tag{3.83}$$

From (3.76a) one notes that $(\nabla' \times \boldsymbol{z}_0)\bar{\varphi} \cdot \boldsymbol{z}_0 = 0$, while from (3.76b), $(\nabla' \times \nabla' \times \boldsymbol{z}_0)\bar{\varphi} \cdot \boldsymbol{z}_0 = -\nabla_t'^2 \bar{\varphi}$. One may write

$$I'(\boldsymbol{r}) = J^0 G'(\boldsymbol{r}, \boldsymbol{r}'), \tag{3.84a}$$

where, in view of $\nabla_t'^2 \phi_i^*(\boldsymbol{\rho}') = -k_{ti}'^2 \phi_i^*(\boldsymbol{\rho}')$ or $\nabla_t^2 \phi_i(\boldsymbol{\rho}) = -k_{ti}'^2 \phi_i(\boldsymbol{\rho})$,

$$\begin{aligned} G'(\boldsymbol{r}, \boldsymbol{r}') &\equiv -\nabla_t'^2 \mathscr{S}'(\boldsymbol{r}, \boldsymbol{r}') = -\nabla_t^2 \mathscr{S}'(\boldsymbol{r}, \boldsymbol{r}') \\ &= \frac{1}{j\omega\varepsilon} \sum_i \phi_i(\boldsymbol{\rho}) \phi_i^*(\boldsymbol{\rho}') Y_i'(z, z'). \end{aligned} \tag{3.84b}$$

Similarly, one writes

$$V''(\boldsymbol{r}) = M^0 G''(\boldsymbol{r}, \boldsymbol{r}'), \tag{3.85a}$$

with

$$\begin{aligned} G''(\boldsymbol{r}, \boldsymbol{r}') &\equiv -\nabla_t'^2 \mathscr{S}''(\boldsymbol{r}, \boldsymbol{r}') = -\nabla_t^2 \mathscr{S}''(\boldsymbol{r}, \boldsymbol{r}') \\ &= \frac{1}{j\omega\mu} \sum_i \psi_i(\boldsymbol{\rho}) \psi_i^*(\boldsymbol{\rho}') Z_i''(z, z'). \end{aligned} \tag{3.85b}$$

One notes from (3.84) and (3.85) that a longitudinal electric current source excites only E modes along z while a longitudinal magnetic current source excites only H modes. The fields are now determined by the following simplified form of (3.79):

$$\boldsymbol{E}(\boldsymbol{r}, \boldsymbol{r}') = \frac{\boldsymbol{J}^0}{j\omega\varepsilon} (\nabla \times \nabla \times \boldsymbol{z}_0) G'(\boldsymbol{r}, \boldsymbol{r}') - M^0 (\nabla \times \boldsymbol{z}_0) G''(\boldsymbol{r}, \boldsymbol{r}'), \tag{3.86a}$$

$$\boldsymbol{H}(\boldsymbol{r}, \boldsymbol{r}') = J^0 (\nabla \times \boldsymbol{z}_0) G'(\boldsymbol{r}, \boldsymbol{r}') + \frac{M^0}{j\omega\mu} (\nabla \times \nabla \times \boldsymbol{z}_0) G''(\boldsymbol{r}, \boldsymbol{r}'). \tag{3.86b}$$

We show now that G' and G'' are scalar Green's functions satisfying, subject to appropriate boundary conditions, the scalar wave equation with an inhomogeneous term $-\delta(\boldsymbol{r} - \boldsymbol{r}')$. Let the operator $(\nabla^2 + k^2)$ act on G' as represented in (3.84b) and assume that the operations of summation and differentiation can be interchanged. Then, since $\nabla_t^2 \phi_t = -k_{ti}'^2 \phi_i$, and $\kappa_i'^2 = k^2 - k_{ti}'^2$,

$$\left(\nabla_t^2 + \frac{\partial^2}{\partial z^2} + k^2\right) G'(\boldsymbol{r}, \boldsymbol{r}') = \frac{1}{j\omega\varepsilon} \sum_i \phi_i(\boldsymbol{\rho}) \phi_i^*(\boldsymbol{\rho}') \left(\frac{d^2}{dz^2} + \kappa_i'^2\right) Y_i'(z, z') \tag{3.87a}$$

$$\begin{aligned} &= -\delta(z - z') \sum_i \phi_i(\boldsymbol{\rho}) \phi_i^*(\boldsymbol{\rho}') \\ &= -\delta(z - z') \delta(\boldsymbol{\rho} - \boldsymbol{\rho}') = -\delta(\boldsymbol{r} - \boldsymbol{r}'). \end{aligned} \tag{3.87b}$$

The transition from (3.87a) to (3.87b) follows via the differential equation for $Y_i'(z,z')$ obtained on elimination of $T_i^V(z,z')$ from (3.62a) and (3.62b), while the identification of the mode function series as $\delta(\boldsymbol{\rho}-\boldsymbol{\rho}')$ is discussed in Section VI of Chapter 2. Thus, the E mode function G' a scalar three-dimensional Green's function which satisfies the inhomogeneous wave equation

$$(\nabla^2+k^2)G'(\boldsymbol{r},\boldsymbol{r}')=-\delta(\boldsymbol{r}-\boldsymbol{r}') \tag{3.88a}$$

subject on the perfectly conducting waveguide boundary s, to the same boundary condition as $\phi_i(\boldsymbol{\rho})$ [see (3.53b)],

$$G'(\boldsymbol{r},\boldsymbol{r}')=0, \qquad \boldsymbol{r} \text{ on } s. \tag{3.88b}$$

The boundary conditions on G' in the z domain will depend on stratification along the z coordinate. For example, across a dielectric interface at $z=z_1$, the transverse electric and magnetic fields are continuous, so the voltage and current in each mode are continuous [see (3.43a) and (3.43b)]. Since $Y_i'(z,z')$ represents a current, continuity of $Y_i'(z,z')$ across z_1 implies from (3.84b) that $G'(\boldsymbol{r},\boldsymbol{r}')$ is likewise continuous across z_1. From the transmission-line equations, the mode voltage is proportional to $(1/\kappa_i'Y_i')(d/dz)Y_i(z,z')$, and since $\kappa_i'Y_i=\omega\varepsilon$, continuity of voltage implies via (3.84b) that $(1/\varepsilon)(\partial/\partial z)G'(\boldsymbol{r},\boldsymbol{r}')$ must likewise be continuous at z_1.[7] Thus, we find that G' and $(1/\varepsilon)(\partial G'/\partial z)$ are required to be continuous across a dielectric interface. Similarly, if the region is terminated at z_1 in a perfectly conducting plane on which the transverse electric field vanishes, each modal voltage vanishes and requires that $\partial G'/\partial z=0$ at z_1, while for an unterminated z domain, a "radiation condition" requiring an outward flow of energy is appropriate. The modal representation for G' in (3.84b) thus constitutes the solution of the Green's function problem posed in (3.88) subject to the above-discussed boundary conditions.

By analogous considerations, one shows that the H mode Green's function G'' in (3.85b) satisfies the inhomogeneous wave equation

$$(\nabla^2+k^2)G''(\boldsymbol{r},\boldsymbol{r}')=-\delta(\boldsymbol{r}-\boldsymbol{r}'), \tag{3.89a}$$

subject on the perfectly conducting waveguide boundary s to the same condition as $\psi_i(\boldsymbol{\rho})$ [see (3.53d)],

$$\frac{\partial G''}{\partial \nu}=0 \qquad \text{on } s. \tag{3.89b}$$

The boundary conditions satisfied by G'' in the z domain are dual to those on G'. At an interface plane $z=z_1$, G'' and $(1/\mu)(\partial G''/\partial z)$ must be continuous, while at a perfectly conducting plane, $G''=0$.[8] The recovery of $\mathscr{S}'$ and $\mathscr{S}''$ from G'

[7] ε and μ in (3.77b), (3.78b), (3.84b), and (3.85b) have constant values appropriate to the medium containing the source point z'; in (3.77), (3.78), (3.80), and (3.86), ε and μ have constant values appropriate to the medium containing the observation point [see also (3.90), (3.92), and (3.94)]. These remarks are relevant for analysis of media with piecewise constant ε and μ.

[8] See the preceding footnote.

and G'', respectively, requires the inversion of (3.84b) and (3.85b). For $k_{ti}^2 \neq 0$, this inversion is accomplished readily in a basis wherein $-\nabla_t^2 \to k_{ti}'^2$ or $k_{ti}''^2$, and leads directly to the representations in (3.77b) and (3.78b).

IX Modal Representations of the Dyadic Green's Functions in an Inhomogeneous Medium

The formulas derived in Section VIII apply to homogeneous media and must be modified if ε and μ are functions of z. In this instance, the results of Sections II.2,V.1, VI, and VII remain valid with the exception of (3.58), which should be written at a source-free point as

$$\boldsymbol{E}(\boldsymbol{r}) = \frac{1}{j\omega\varepsilon(z)}(\nabla \times \nabla \times \boldsymbol{z}_0)I'(\boldsymbol{r}) - (\nabla \times \boldsymbol{z}_0)V''(\boldsymbol{r}), \tag{3.90a}$$

$$\boldsymbol{H}(\boldsymbol{r}) = \frac{1}{j\omega\mu(z)}(\nabla \times \nabla \times \boldsymbol{z}_0)V''(\boldsymbol{r}) + (\nabla \times \boldsymbol{z}_0)I'(\boldsymbol{r}) \tag{3.90b}$$

with $I'(\boldsymbol{r})$ and $V''(\boldsymbol{r})$ defined in (3.57). As regards the results in Section VIII, one notes from the method of derivation that (3.72)–(3.76) still apply provided that ε and μ are replaced by $\varepsilon(z')$ and $\mu(z')$, respectively. It then follows that (3.77) should be written as

$$I'(\boldsymbol{r}) = -L_1'\mathscr{S}_d' \cdot \boldsymbol{M}^0 + \frac{1}{j\omega\varepsilon(z')}L_2'\mathscr{S}_d' \cdot \boldsymbol{J}^0, \tag{3.91a}$$

where the vector operators L_1' and L_2' are defined as

$$L_1' \equiv \nabla' \times \boldsymbol{z}_0, \quad L_2' \equiv \nabla' \times \nabla' \times \boldsymbol{z}_0, \tag{3.91b}$$

and

$$\mathscr{S}_d' = \sum_i \frac{\phi_i(\boldsymbol{\rho})\phi_i^*(\boldsymbol{\rho}')}{k_{ti}'^2} Y_i'(z, z'). \tag{3.91c}$$

Dual considerations apply to (3.78).

With the above modifications, the dyadic Green's functions in (3.80) are now written in the following form:

$$\mathscr{Z}(\boldsymbol{r}, \boldsymbol{r}') = \frac{1}{\omega^2\varepsilon(z)\varepsilon(z')}L_2L_2'\mathscr{S}_d' + L_1L_1'\mathscr{S}_d'', \tag{3.92a}$$

$$\mathscr{Y}(\boldsymbol{r}, \boldsymbol{r}') = \frac{1}{\omega^2\mu(z)\mu(z')}L_2L_2'\mathscr{S}_d'' + L_1L_1'\mathscr{S}_d', \tag{3.92b}$$

$$\mathscr{T}_e(\boldsymbol{r}, \boldsymbol{r}') = \frac{1}{j\omega\varepsilon(z)}L_2L_1'\mathscr{S}_d' + \frac{1}{j\omega\mu(z')}L_1L_2'\mathscr{S}_d'', \tag{3.92c}$$

$$-\mathscr{T}_m(\boldsymbol{r}, \boldsymbol{r}') = \frac{1}{j\omega\mu(z)}L_2L_1'\mathscr{S}_d'' + \frac{1}{j\omega\varepsilon(z')}L_1L_2'\mathscr{S}_d', \tag{3.92d}$$

where

$$L_1 \equiv \nabla \times \boldsymbol{z}_0, \quad L_2 \equiv \nabla \times \nabla \times \boldsymbol{z}_0, \quad \mathscr{S}''_d = \sum_i \frac{\psi_i(\boldsymbol{\rho})\psi_i^*(\boldsymbol{\rho})}{k''^2_{ti}} Z''_i(z, z'). \quad (3.92e)$$

It is readily verified that these more general expressions satisfy, as they must, the reciprocity relations (3.82).

The modal Green's functions $Y'_i(z, z')$ and $Z''_i(z, z')$ are defined in (3.62). Because $\kappa(z) = [\omega^2 \mu(z)\varepsilon(z) - k^2_{ti}]^{1/2}$ is now variable, the characteristic impedances $Z_i(z)$ and admittances $Y_i(z)$ are also functions of z, so the associated transmission lines are non-uniform.[9] On elimination of T_i^V and T_i^I from (3.62a), (3.62b) and (3.62c), (3.62d), respectively, one finds that the modal Green's functions satisfy the following second-order differential equations [note from (3.51c, d) that $\kappa'_i(z)Y'_i(z) = \omega\varepsilon(z), \kappa''_i(z)Z''_i(z) = \omega\mu(z)$]:

$$[D^2_\varepsilon(z) + \kappa'^2_i(z)]Y'_i(z, z') = -j\omega\varepsilon(z')\delta(z - z'), \quad (3.93a)$$

$$[D^2_\mu(z) + \kappa''^2_i(z)]Z''_i(z, z') = -j\omega\mu(z')\delta(z - z'), \quad (3.93b)$$

where

$$D^2_\alpha(z) = \alpha(z)\frac{d}{dz}\frac{1}{\alpha(z)}\frac{d}{dz}, \qquad \alpha = \varepsilon \text{ or } \mu. \quad (3.93c)$$

The boundary conditions at the endpoints of the transmission line are phrased as in (3.65). Note that the E mode terminal impedance is given via (3.62a) and (3.62b) by $[(d/dz)Y'_i(z, z')/ - j\kappa'_i Y'_i Y'_i(z, z')]_{z1,2}$; the spatially varying characteristic impedance here should not be confused with the terminal impedance in Section VII. At a junction between two transmission lines with parameters $\kappa_{i1}(z)$, $Z_{i1}(z)$ and $\kappa_{i2}(z)$, $Z_{i2}(z)$, respectively, the voltage and current are continuous. Thus, from (3.62), $Y'_i(z, z')$, $[1/\varepsilon(z)](d/dz)Y'_i(z, z')$, and $Z''_i(z, z')$, $[1/\mu(z)](d/dz)Z''_i(z, z')$ are continuous across the junction point.

If the sources are longitudinal, (3.92) simplify and lead to expressions analogous to those in Section VIII. In fact, one obtains expressions similar to (3.86):

$$\boldsymbol{E}(\boldsymbol{r}, \boldsymbol{r}') = \frac{J^0}{j\omega\varepsilon(z)} L_2 G'(\boldsymbol{r}, \boldsymbol{r}') - M^0 L_1 G''(\boldsymbol{r}, \boldsymbol{r}'), \quad (3.94a)$$

$$\boldsymbol{H}(\boldsymbol{r}, \boldsymbol{r}') = J^0 L_1 G'(\boldsymbol{r}, \boldsymbol{r}') + \frac{M^0}{j\omega\mu(z)} L_2 G''(\boldsymbol{r}, \boldsymbol{r}'), \quad (3.94b)$$

where

[9] Although the waveguide region is geometrically uniform in that successive geometrical cross sections transverse to z are identical, an electrical non-uniformity is introduced by the longitudinal variability of the medium constants. Consequently, the network representation involves non-uniform transmission lines representative of the z behavior of a typical mode.

$$G'(\boldsymbol{r},\boldsymbol{r}') = \frac{1}{j\omega\varepsilon(z)}\sum_i Y_i'(z,z')\phi_i(\boldsymbol{\rho})\phi_i^*(\boldsymbol{\rho}') = -\frac{1}{j\omega\varepsilon(z')}\nabla_t'^2\mathscr{S}_d', \tag{3.95a}$$

$$G''(\boldsymbol{r},\boldsymbol{r}') = \frac{1}{j\omega\mu(z')}\sum_i Z_i''(z,z')\psi_i(\boldsymbol{\rho})\psi_i^*(\boldsymbol{\rho}') = -\frac{1}{j\omega\mu(z')}\nabla_t'^2\mathscr{S}_d''. \tag{3.95b}$$

The differential equations for the scalar Green's functions G' and G'' are now in view of (3.93):

$$[\mathscr{D}_\varepsilon^2(z) + \nabla_t^2 + k^2(z)]G'(\boldsymbol{r},\boldsymbol{r}') = -\delta(\boldsymbol{r}-\boldsymbol{r}'), \quad k^2(z) = \omega^2\mu(z)\varepsilon(z), \tag{3.96a}$$

$$[\mathscr{D}_\mu^2(z) + \nabla_t^2 + k^2(z)]G''(\boldsymbol{r},\boldsymbol{r}') = -\delta(\boldsymbol{r}-\boldsymbol{r}'), \tag{3.96b}$$

where

$$\mathscr{D}_\alpha^2(z) = \alpha(z)\frac{\partial}{\partial z}\frac{1}{\alpha(z)}\frac{\partial}{\partial z}. \tag{3.96c}$$

It may also be verified that the Green's function $G'(\boldsymbol{r},\boldsymbol{r}')/\sqrt{\varepsilon(z)}$ satisfies the wave equation with the modified wavenumber $\bar{k}(z)$:

$$[\nabla^2 + \bar{k}^2(z)]\frac{G'(\boldsymbol{r},\boldsymbol{r}')}{\sqrt{\varepsilon(z)}} = -\frac{\delta(\boldsymbol{r}-\boldsymbol{r}')}{\sqrt{\varepsilon(z')}}, \quad \bar{k}^2(z) = k^2(z) - \sqrt{\varepsilon(z)}\frac{d^2}{dz^2}\frac{1}{\sqrt{\varepsilon(z)}} \tag{3.97}$$

with a dual relation applicable to $G''(\boldsymbol{r},\boldsymbol{r}')/\sqrt{\mu(z)}$. Corresponding equations for $\mathscr{S}_d'$ and $\mathscr{S}_d''$ follow on use of (3.95). The conditions satisfied by G' and G'' on the transverse and longitudinal boundaries of the region are the same as those deduced in connection with (3.88) and (3.89). These boundary conditions, in conjunction with (3.96), render the specification of G' and G'' unique. The modal representations in (3.95) constitute solutions for G' and G'' and are directly deducible from a z-transmission analysis. Alternative representations of the solution for G' and G'' can also be constructed.
All the above relations reduce to those in Section VIII when ε and μ are constant.

X Network–Oriented Formulation of the Characteristic Green's Functions

In the Sturm-Liouville (SL) problems discussed in Sect. VI of Chapter 2, which culminated with the formulation of the SL eigenvalue problem via characteristic Green's functions (Chapter 2, Section VI.3), the Green's functions (GFs) could be taken to represent any generic scalar field variable. Because of the emphasis in this volume on the connection between *fields* and *networks*, it is appropriate to relate the generic GFs to the source-excited modal voltages and current GFs used in network analysis. These V_i, I_i GFs are defined by, and propagate (along the rectilinear coordinate z) according to the modal transmission line equations (3.51a-3.51d), where $u \to z$ represents a rectilinear coordinate.

For the general case where the ambient medium has a z-dependent permittivity $\varepsilon(z)$ and permeability $\mu(z)$, the second-order SL type differential equations result on elimination of either the V_i or I_i from (3.51a, 3.51b) with (3.51c, 3.51d). If the current GF I_i is eliminated via (3.51a), one sets $v_i \equiv 0$, $i_i = i_i(z') = \delta(z - z')$, i.e.,

$$-\frac{dV(z,z')}{dz} = jk_z(z)Z(z)I(z,z'), \tag{3.98a}$$

$$-\frac{dI(z,z')}{dz} = jk_z(z)Y(z)V(z,z') - \delta(z-z'), \tag{3.98b}$$

where $Z(z) = 1/Y(z)$ and $k_z(z)$ are the characteristic impedance and propagation constant. For an H mode transmission line with distinguishing double-prime superscripts one has $k_z''\, Z'' = \omega\mu$, (see (3.51d)) whence this format is preferred for the H modes. Thus, the corresponding SL equation for $V''(z,z')$ has the form (2.127, 2.128):

$$\left\{\frac{d}{dz}\left[\frac{1}{\bar{\mu}(z)}\frac{d}{dz}\right] + k_0^2\bar{\varepsilon}(z) - \frac{k_T''^2}{\bar{\mu}(z)}\right\} V''(z,z') = -j\omega\mu_0\delta(z-z'), \tag{3.99a}$$

where $k_0^2 = \omega^2\mu_0\varepsilon_0$ and

$$\bar{\mu}(z) = \frac{\mu(z)}{\mu_0}, \qquad \bar{\varepsilon}(z) = \frac{\varepsilon(z)}{\varepsilon_0}, \tag{3.99b}$$

μ_0 and ε_0 representing convenient reference values for the thus normalized permeability and permittivity, respectively. Upon comparing (2.127, 2.128) and (3.99a) one makes the identifications:

$$p(z) = w(z) = \frac{1}{\bar{\mu}(z)}, \quad q(z) = -k_0^2\bar{\varepsilon}(z), \quad k_T''^2 = -\lambda,$$

$$V''(z,z') = j\omega\mu_0 g_z''(z,z';\lambda_z). \tag{3.100}$$

The H mode current Green's functions (GF) is now obtained from (3.98a). The boundary conditions in (2.129) become via (3.98a) and (3.100)

$$\frac{I''}{V''} = j\frac{p(dg''/dz)}{\omega\mu_0 g''}, \tag{3.101a}$$

and defining the terminating admittances $\overrightarrow{Y}_T''$ and $\overleftarrow{Y}_T''$ at z_1 and z_2 (as looking toward the terminations):

$$\overleftarrow{Y}_T'' = -\frac{I''(z_1,z')}{V''(z_1,z')} = \frac{j\gamma_1}{\omega\mu_0}, \qquad \overrightarrow{Y}_T'' = \frac{I''(z_2,z')}{V''(z_2,z')} = \frac{j\gamma_2}{\omega\mu_0}. \tag{3.101b}$$

Concerning the behavior of $V_i''(z,z')$ and $I_i''(z,z')$ across the point current source at $z = z'$, one observes from (3.98a) that V_i'' is continuous at z', because its derivative there is bounded. On the other hand, from (3.98b), the derivative of I_i'' gives rise to the delta function $\delta(z-z')$, implying that I_i'' has a jump discontinuity across $z = z'$. Thus,

$$V''(z,z')\big|_{z'-\Delta}^{z'+\Delta} = 0, \qquad \Delta \longrightarrow 0, \tag{3.102a}$$

while the discontinuity in the current is given by

$$I''(z,z')\big|_{z'-\Delta}^{z'+\Delta} = 1. \tag{3.102b}$$

and the corresponding conditions on g are

$$g_z''(z,z';\lambda_z)\big|_{z'-\Delta}^{z'+\Delta} = 0, \qquad p(z)\frac{d}{dz}g_z''(z,z';\lambda_z)\big|_{z'-\Delta}^{z'+\Delta} = -1. \tag{3.102c}$$

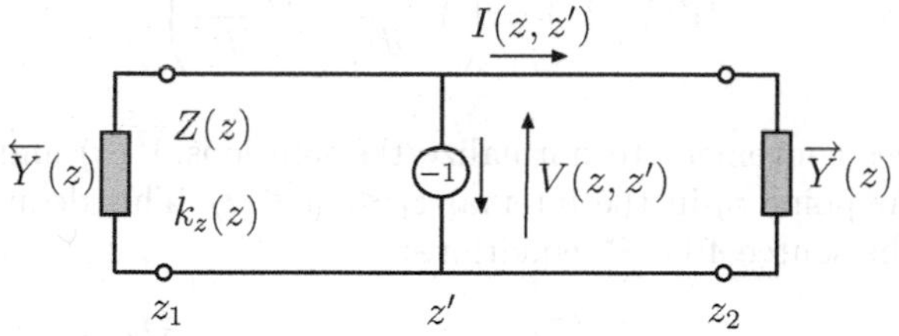

Fig. 3.2. Non-uniform modal transmission line excited with a unit current generator.

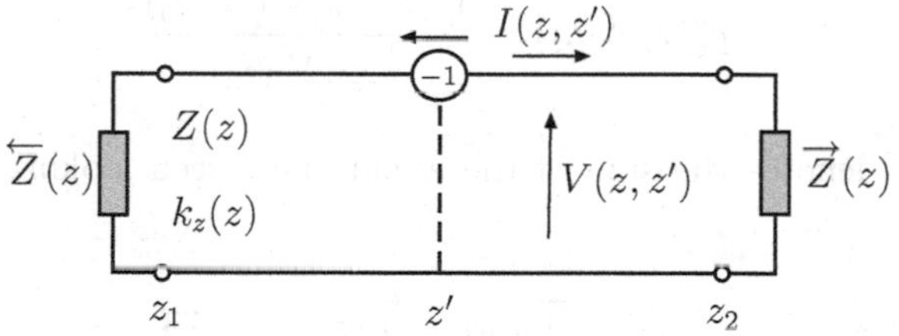

Fig. 3.3. Non-uniform modal transmission line excited with a unit voltage generator.

The network schematization of these relations is shown in Figure 3.2.
By considerations dual to those employed above, one notes from (3.51c) that $\kappa_i(z)Y_i(z) = \omega\varepsilon(z)$ whence this property favors evaluation of the E mode current GF $I_i(z,z')$, defined via (3.51a),(3.51b) with $i_i'(z) \equiv 0$. The resulting E mode equations can be written down directly by making the following duality replacements in (3.99a)–(3.102):

$$V'' \to I'', \quad I'' \to V', \quad \bar{\mu} \leftrightarrow \bar{\varepsilon}, \quad \mu_0 \leftrightarrow \varepsilon_0, \quad k''_T \to k'_T, \quad Y''_T \to Z'_T, \quad g'' \to g'. \tag{3.103}$$

The corresponding network schematization is shown in Figure 3.3. The construction of the voltage and current Green's functions can be performed directly from Sect. (VI.2) of Chapter 2. For the H mode Green's function (GF) $g''_i(z,z')$ we replace the functions $\overleftarrow{f}(z)$ and $\overrightarrow{f}(z)$ by the functions $\overleftarrow{V}(z)$ and $\overrightarrow{V}(z)$, respectively. Both sets of functions satisfy the source–free Sturm-Louiville (SL) equations as well as the boundary conditions at z_1 and z_2, respectively. It then follows from (2.187) that,

$$g''_z(z,z';\lambda_z) = \frac{\overleftarrow{V}(z_<)\overrightarrow{V}(z_>)}{-pW(\overleftarrow{V},\overrightarrow{V})}, \tag{3.104a}$$

with the Wronskian given by,

$$W(\overleftarrow{V},\overrightarrow{V}) = \left(\overleftarrow{V}\frac{d\overrightarrow{V}}{dx} - \overrightarrow{V}\frac{d\overleftarrow{V}}{dx}\right). \tag{3.104b}$$

It is sometimes convenient to normalize the solutions $\overleftarrow{V}(z)$ and $\overrightarrow{V}(z)$ to unity at a particular point z_0 in the interval $z_1 \le z_0 \le z_2$. This defines the following solutions of the source–free SL equations:

$$\overleftarrow{V}(z,z_0) = \frac{\overleftarrow{V}(z)}{\overleftarrow{V}(z_0)}, \qquad \overrightarrow{V}(z,z_0) = \frac{\overrightarrow{V}(z)}{\overrightarrow{V}(z_0)}, \tag{3.105}$$

with the corresponding Green's function solution

$$g''_z(z,z';\lambda_z) = \frac{\overleftarrow{V}(z_<,z_0)\overrightarrow{V}(z_>,z_0)}{j\omega\mu_0 \overleftrightarrow{Y}(z_0)}, \tag{3.106a}$$

where $\overleftrightarrow{Y}(z_0)$ denotes the sum of the admittances seen looking to the left and right from z_0:

$$\overleftrightarrow{Y}(z_0) = \overrightarrow{Y}(z_0) + \overleftarrow{Y}(z_0) = \frac{\overrightarrow{I}(z_0)}{\overrightarrow{V}(z_0)} + \frac{\overleftarrow{I}(z_0)}{\overleftarrow{V}(z_0)}, \qquad \overleftrightarrows{I}(z_0) = \pm\frac{p}{j\omega\mu_0}\left.\frac{\overleftrightarrows{V}}{dz}\right|_{z_0}. \tag{3.106b}$$

Note that all the functions on the right-hand side of (3.106b) are λ–dependent. The construction of the modal completeness relation (delta function representation) via the characteristic Green's function method can be performed for the network-oriented GFs, yielding for the H mode problem (upon exhibiting the λ–dependence),

$$\bar{\mu}(z)\delta(z-z') = -\frac{1}{2\pi j}\oint_C g_z''(z,z';\lambda_z)\,d\lambda = \sum_\alpha \hat{\psi}_\alpha(z)\hat{\psi}_\alpha^*(z') \tag{3.107a}$$

$$= -\frac{1}{2\pi j}\oint_C \frac{\overleftarrow{V}(z_<,z_0;\lambda)\overrightarrow{V}(z_>,z_0;\lambda)}{j\omega\mu_0\overleftrightarrow{Y}(z_0,\lambda)}\,d\lambda \tag{3.107b}$$

$$= \sum_\alpha \frac{\overleftarrow{V}(z_<,z_0;\lambda_\alpha)\overleftarrow{V}^*(z_>,z_0;\lambda_\alpha)}{-j\omega\mu_0(\partial/\partial\lambda_\alpha)\overleftrightarrow{Y}(z_0,\lambda_\alpha)}\frac{1}{2\pi j}\oint_C \frac{d\lambda}{\lambda-\lambda_\alpha} \tag{3.107c}$$

$$= \sum_\alpha \frac{\overleftarrow{V}(z,z_0;\lambda_\alpha)\overleftarrow{V}^*(z',z_0;\lambda_\alpha)}{\omega\mu_0(\partial/\partial\lambda_\alpha)\overleftrightarrow{B}(z_0,\lambda_\alpha)}, \qquad \overleftrightarrow{Y} = j\overleftrightarrow{B}. \tag{3.107d}$$

The resonant condition determining the eigenvalues λ_α (poles in the complex λ plane) is given by

$$\overleftrightarrow{Y}(z_0,\lambda_\alpha) = 0. \tag{3.107e}$$

The normalized mode functions $\hat{\psi}_\alpha^*(z)$ are therefore given by

$$\hat{\psi}_\alpha = \frac{1}{\sqrt{\omega\mu_0(\partial/\partial\lambda_\alpha)\overleftrightarrow{B}(z_0,\lambda_\alpha)}}\overleftarrow{V}(z,z_0;\lambda_\alpha). \tag{3.107f}$$

A typical contour of integration in the complex λ plane is sketched in Figure 3.4.

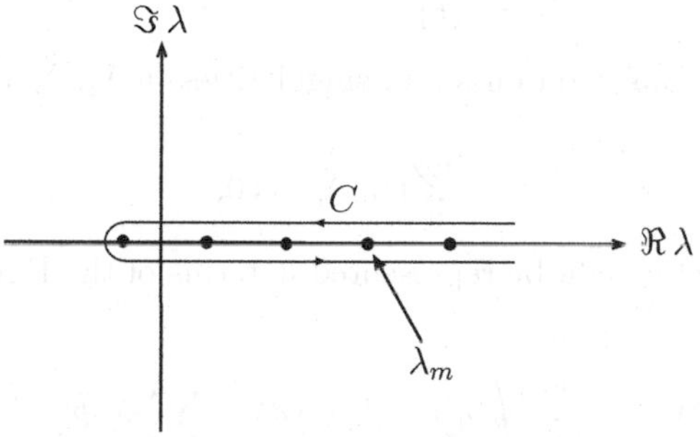

Fig. 3.4. Contour of integration.

Again, as before, the corresponding constructions for the E mode Green's functions can be carried out in a similar (dual) fashion via (3.103). Thus, the E mode characteristic Green's function $g_z'(z,z';\lambda_z)$ is given by:

$$g_z'(z,z';\lambda_z) = \frac{\overleftarrow{I}(z_<,z_0)\overrightarrow{I}(z_>,z_0)}{j\omega\varepsilon_0\overleftrightarrow{Z}(z_0)}, \qquad \overleftrightarrow{Z}(z_0) = \overleftarrow{Z}(z_0) + \overrightarrow{Z}(z_0), \tag{3.108}$$

where the primes, distinctive of the E mode problem, have been omitted from $\overleftrightarrow{I}$ and the total impedance function $\overleftrightarrow{Z}(x_0)$. The eigenvalues λ_α are specified implicitly by the resonance equation

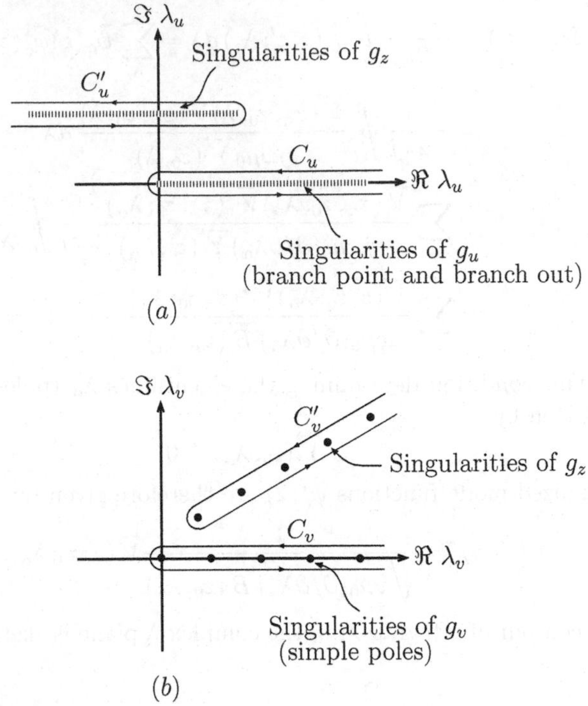

Fig. 3.5. Contours and singularities in λ_u, λ_v planes.

$$\overleftrightarrow{Z}(z_0, \lambda_\alpha) = 0, \tag{3.109}$$

and the delta function can be represented in terms of the E mode eigenfunctions $\hat{\Phi}_\alpha$ as

$$\begin{aligned}\bar{\varepsilon}(z)\delta(z - z') &= -\frac{1}{2\pi j}\oint_C g'_z(z, z'; \lambda_z)\, d\lambda = \sum_\alpha \hat{\Phi}_\alpha \hat{\Phi}^*_\alpha(z') \\ &= \sum_\alpha \frac{\overleftarrow{I}(z, z_0; \lambda_\alpha)\overleftarrow{I}^*(z, z_0, \lambda_\alpha)}{\omega\varepsilon_0(\partial/\partial\lambda)\overleftrightarrow{X}(x'_0, \lambda_m)}, \qquad \overleftrightarrow{Z} = j\overleftrightarrow{X}.\end{aligned} \tag{3.110}$$

Thus, the discrete orthonormal E mode eigenfunctions $\hat{\Phi}_\alpha$ are given by

$$\hat{\Phi}_\alpha(z) = \frac{1}{\sqrt{\omega\varepsilon_0(\partial/\partial\lambda_\alpha)\overleftrightarrow{X}(z_0, \lambda_\alpha)}}\overleftarrow{I}(z, z_0; \lambda), \qquad z_1 \le z \le z_2. \tag{3.110a}$$

An alternative approach is based on modal reflection coefficients instead of modal impedances. The transmission line relations for this z-dependent medium are

$$V(z) = V_+(z) + V_-(z) = V_+(1 + \overrightarrow{\Gamma}_V) = V_-(1 + \overleftarrow{\Gamma}_V), \tag{3.111a}$$

$$I(z) = I_+(z) + I_-(z) = I_+(1 + \overrightarrow{\Gamma}_I) = I_-(1 + \overleftarrow{\Gamma}_I), \tag{3.111b}$$

where the subscripts $_+$ and $_-$ on V or I denote wave components traveling in the $+z$ and $-z$ directions, respectively, and $\overleftrightarrows{\Gamma}_V$ ($\overleftrightarrows{\Gamma}_I$) are the voltage (current) reflection coefficients seen when looking along the $\pm z$ directions:

$$\overleftrightarrows{\Gamma}_V = \frac{V_\mp}{V_\pm}, \qquad \overleftrightarrows{\Gamma}_I = \frac{I_\mp}{I_\pm}. \tag{3.111c}$$

If $\overrightarrow{\zeta} = V_+/I_+$ denotes the input impedance of a matched transmission line looking in the $+z$ direction, and $\overleftarrow{\zeta} = -V_-/I_-$ represents the input impedance in the $-z$ direction, then

$$\overrightarrow{\Gamma}_I = -\frac{\overrightarrow{\zeta}}{\overleftarrow{\zeta}}\overrightarrow{\Gamma}_V, \qquad \overleftarrow{\Gamma}_I = -\frac{\overleftarrow{\zeta}}{\overrightarrow{\zeta}}\overleftarrow{\Gamma}_V, \tag{3.112a}$$

and

$$\overrightarrow{Z}(z) = \overrightarrow{\zeta}(z)\frac{1 + \overrightarrow{\Gamma}_V(z)}{1 - \frac{\overrightarrow{\zeta}(z)}{\overleftarrow{\zeta}(z)}\overrightarrow{\Gamma}_V(z)}, \qquad \overleftarrow{Z}(z) = \overleftarrow{\zeta}(z)\frac{1 + \overleftarrow{\Gamma}_V(z)}{1 - \frac{\overleftarrow{\zeta}(z)}{\overrightarrow{\zeta}(z)}\overleftarrow{\Gamma}_V(z)}. \tag{3.112b}$$

Conversely,

$$\overrightarrow{\Gamma}_V(z) = \frac{\frac{\overrightarrow{Z}(z)}{\overrightarrow{\zeta}(z)} - 1}{\frac{\overrightarrow{Z}(z)}{\overleftarrow{\zeta}(z)} + 1}, \qquad \overleftarrow{\Gamma}_V(z) = \frac{\frac{\overleftarrow{Z}(z)}{\overleftarrow{\zeta}(z)} - 1}{\frac{\overleftarrow{Z}(z)}{\overrightarrow{\zeta}(z)} + 1}. \tag{3.112c}$$

The transverse resonance relation

$$\overrightarrow{Z}(z) + \overleftarrow{Z}(z) = 0$$

becomes

$$\overrightarrow{\Gamma}_V(z)\overleftarrow{\Gamma}_V(z) = 1 = \overrightarrow{\Gamma}_I(z)\overleftarrow{\Gamma}_I(z). \tag{3.113}$$

The above traveling-wave formulation leads to a set of eigenfunctions alternative to that in (3.110).

X.1 Alternative Representations

The theory of alternative multidimensional Green's function representations is based on use of the one-dimensional characteristic Green's functions. For uniform waveguide regions describable in a $(\boldsymbol{\rho}, z)$ coordinate system the two-dimensional eigenfunctions $\Phi_i(\boldsymbol{\rho})$ are of the form

$$\Phi_i(\boldsymbol{\rho}) = \Phi_\alpha(u)\Phi_\beta(v), \qquad \boldsymbol{\rho} = (u, v), \tag{3.114}$$

where $\Phi_\alpha(u)$ and $\Phi_\beta(v)$ are one-dimensional orthonormal functions in separable u and v coordinate spaces transverse to z. The two-dimensional completeness relation involving $\Phi_i(\boldsymbol{\rho})$ is:

$$\delta(\boldsymbol{\rho}-\boldsymbol{\rho}') = \frac{\delta(u-u')\delta(v-v')}{h_u h_v} = \sum_i \Phi_i(\boldsymbol{\rho})\Phi_i^*(\boldsymbol{\rho}') \tag{3.115a}$$

$$= \sum_\alpha \Phi_\alpha(u)\Phi_\alpha^*(u') \sum_\beta \Phi_\beta(v)\Phi_\beta^*(v'), \tag{3.115b}$$

where the curvilinear metric parameters h_u and h_v in (3.115a) are defined via the relation $dS = h_u h_v\, du\, dv$, and dS is an area element in the cross section. Then from the above equation applied to the u-dependent functions,

$$\frac{\delta(u-u')}{h_u} = \sum_\alpha \Phi_\alpha(u)\Phi_\alpha^*(u') = \frac{1}{2\pi j}\sum_\alpha \oint_{C_u} \frac{\Phi_{\lambda_u}(u)\Phi_{\lambda_u}^*(u')}{\lambda_u - \lambda_\alpha}\, d\lambda_u$$

$$= -\frac{1}{2\pi j}\oint_{C_u} g_u(u,u';\lambda_u)\, d\lambda_u, \tag{3.116}$$

where $\Phi_{\lambda_\alpha} \equiv \Phi_\alpha$, g_u is the characteristic Green's function associated with the eigenvalue problem in the u domain, and the contour C_u in the complex λ_u plane encloses in the positive sense all the singularities (poles or branch points, with associated branch cuts) of g_u. The less general first representation in (3.116), involving the discrete or continuous sum over the eigenvalues λ_α is obtained by evaluating the contour integral in terms of the singularities of g_u. The analogue of (3.116) for the v domain is

$$\frac{\delta(v-v')}{h_v} = \sum_\beta \Phi_\beta(v)\Phi_\beta^*(v') = -\frac{1}{2\pi j}\oint_{C_v} g_v(v,v';\lambda_v)\, d\lambda_v \tag{3.117}$$

with C_v defined similarly to C_u leading to the most general, two-dimensional completeness relation

$$\delta(\boldsymbol{\rho}-\boldsymbol{\rho}') = \Big[-\frac{1}{2\pi j}\oint_{C_u} g_u(u,u';\lambda_u)\, d\lambda_u\Big]\Big[-\frac{1}{2\pi j}\oint_{C_v} g_v(v,v';\lambda_v)\, d\lambda_v\Big] \tag{3.118a}$$

$$= \frac{1}{(-2\pi j)^2}\oint_{C_u}\oint_{C_v} g_u(u,u';\lambda_u) g_v(v,v';\lambda_v)\, d\lambda_u\, d\lambda_v. \tag{3.118b}$$

When the eigenfunctions in (3.116) or (3.117) are used to represent a three-dimensional Green's function in (u,v,z) space, one obtains

$$G(\boldsymbol{r},\boldsymbol{r}') = \sum_i \Phi_i(\boldsymbol{\rho})\Phi_i^*(\boldsymbol{\rho}') g_z(z,z';\lambda_{zi}). \tag{3.119}$$

The z-dependent modal Green's function g_z satisfies a one-dimensional equation obtained after elimination of the (u,v) dependence from the corresponding three-dimensional equation via (3.115a) and (3.115b). On comparing (3.115), (3.118a), and (3.119), one notes that the three-dimensional scalar Green's function G can be represented in terms of the one-dimensional characteristic Green's functions[10] g_u, and g_v, and the modal Green's function g_z, as follows:

$$G(\boldsymbol{r}, \boldsymbol{r}') = \frac{1}{(-2\pi j)^2} \oint_{C_u} \oint_{C_v} g_u(u, u'; \lambda_u) g_v(v, v'; \lambda_v) g_z(z, z'; \lambda_z)\, d\lambda_v\, d\lambda_u. \quad (3.120)$$

The contour C_u in the complex λ_u plane encloses in the positive sense all singularities of g_u *but no others*, while the contour C_v in the complex λ_v plane encloses in the positive sense all singularities of g_v *but no others*. Additional singularities in the λ_u and (or) λ_v planes arise due to $g_z(z, z'; \lambda_z)$; it is recognized that generally $\lambda_z = \lambda_z(\lambda_u, \lambda_v)$, where the detailed dependence of λ_z on λ_u and λ_v is dictated by the particular coordinate representation in the u, v domain. For example, with $\lambda_z \equiv \kappa_i'^2 = k^2 - k_{ti}'^2$, we have

$$\lambda_z = k^2 - \lambda_u - \lambda_v \qquad \text{for rectangular coordinates } u \equiv x,\ v \equiv y, \quad (3.121a)$$

whereas in cylindrical coordinates, with $k_{ti}'^2 \equiv p^2 \to \lambda_u$,

$$\lambda_z = k^2 - \lambda_u \qquad \text{for cylindrical coordinates } u \equiv \rho,\ v \equiv \varphi^{11}. \quad (3.121b)$$

The contour integral representation in (3.120), involving the one-dimensional Green's functions g_u, g_v, and g_z, can be considered as the most general separable representation for the three-dimensional Green's function G. Upon evaluating the contour integrals in (3.120) in terms of the discrete and (or) continuous spectra arising from the pole or branch-cut singularities, respectively, of g_u and g_v, and noting that g_z has no singularities inside the contours C_u and C_v, one recovers the original z-transmission formulation in (3.119). Different representations are also obtainable by contour deformations in the λ_u and λ_v planes. Typical examples wherein g_u, g_v, and g_z have singularities in the λ_u and λ_v planes are shown in Figure 3.5. The functions g_u, g_v, and g_z are so defined as to vanish sufficiently rapidly at infinity in the λ_u and λ_v planes. This is achieved by an appropriate choice of branch cuts on Riemann surfaces, associated with any existing branch-point singularities of the g functions, so as to result in negligeable contributions to the integral in (3.120) from closed contours as $|\lambda_u| \to \infty$ and $|\lambda_v| \to \infty$. The path C_u in Figure 3.5a can therefore be deformed into the path C_u' enclosing the singularities of g_z in the λ_u plane, to yield

---

[10] As pointed out in Section 3.3a of reference [14], the modal and characteristic Green's functions differ only in that the parameter λ is specified for the former ($\lambda = \lambda_i$), but unspecified for the latter.

[11] In this case, $g_u \equiv g_\rho$ depends also on λ_v, so one should write $g_u \to g_u(u, u'; \lambda_u, \lambda_v)$. Thus, g_u has singularities in both the λ_u and λ_v planes, while g_z has singularities in the λ_u plane only. Only the singularities of g_u enclosed by the contour C_u in the complex λ_u plane contribute to the modal representation for G as in (3.119).

$$G(\boldsymbol{r},\boldsymbol{r}') = \frac{1}{(-2\pi j)^2} \oint_{C'_u} \oint_{C_v} g_u(u,u';\lambda_u) g_v(v,v';\lambda_v) g_z(z,z';\lambda_z)\, d\lambda_v\, d\lambda_u \quad (3.122a)$$

$$= \sum_\beta \Phi_\beta(v)\Phi^*_\beta(v') \sum_\gamma \Phi_\gamma(z)\Phi^*_\gamma(z')\lambda_{u\gamma\beta}, \quad (3.122b)$$

where the modal representation in (3.122b) is obtained upon evaluating the integrals over the contours C'_u and C_v in (3.122a). The $\Phi_\gamma(z)$ denote the eigenfunctions in the z-domain arising from the eigenvalue problem associated with g_z, λ_u being the characteristic parameter.[12] In (3.122a), g_v and g_z are now characteristic Green's functions, while g_u is a modal Green's function wherein λ_u takes on the values specified along C'_u. Because of the explicit presence of $g_u(u,u';\lambda_{ur\beta})$ in (3.122b), one identifies this representation as arising from a guided-wave analysis in which the transmission direction is taken along the u coordinate.
Alternatively, one may deform the contour C_v into the contour C'_v in the complex λ_v plane as shown in Figure 3.5(b) to obtain

$$G(\boldsymbol{r},\boldsymbol{r}') = \frac{1}{(-2\pi j)^2} \oint_{C_u} \oint_{C'_v} g_u(u,u';\lambda_u) g_v(v,v';\lambda_v) g_z(z,z';\lambda_z)\, d\lambda_v\, d\lambda_u, \quad (3.123a)$$

$$= \sum_s \Phi_s(z)\Phi^*_s(z') \sum_\alpha \Phi_\alpha(u)\Phi^*_\alpha(u') g'_v(v,v';\lambda_{vs\alpha}). \quad (3.123b)$$

The modal representation in (3.123b) is derived by considerations analogous to the above and is identified as a v-transmission formulation. The $\Phi_s(z)$ are the eigenfunctions in the z domain arising from the eigenvalue problem associated with g_z as the characteristic Green's function and λ_v as the characteristic parameter. Additional representations are possible wherein, for example, only the integral C_u in (3.123a) is evaluated in terms of the mode spectrum in u while the integral C'_v remains unchanged. It is to be emphasized that all of the above alternative representations are to be considered as formal in that the deformability of contours must be verified in each case.
For a radial transmission formulation, as in (3.121b), g_z is not a function of λ_v; instead, g_u is a function of both λ_u and λ_v. Now, the contour C'_v encloses the singularities of g_u in the λ_v plane, with λ_u. treated as a fixed parameter. Moreover, one notes that

$$\left(\frac{d}{d\rho}\rho\frac{d}{d\rho} + \lambda_u\rho - \frac{\lambda_v}{\rho}\right) g_u(\rho,\rho';\lambda_u,\lambda_v) = -\delta(\rho-\rho'), \quad (3.124)$$

whence instead of (3.123a),

[12] For non-Hermitian problems with complex eigenvalues, the spectral representation involves the symmetric form wherein $\Phi^*_\gamma(z')$ is replaced by $\Phi_\gamma(z)$, or more generally by an "adjoint" function $\bar{\Phi}_\gamma(z')$.

$$G(\boldsymbol{r},\boldsymbol{r}') = \frac{1}{(-2\pi j)^2} \oint_{C_u'} \oint_{C_v'} g_u'(u,u';\lambda_u,\lambda_v) g_v'(v,v';\lambda_v) g_z'(z,z';k^2-\lambda_u)\, d\lambda_v\, d\lambda_u. \tag{3.125}$$

Equation (3.123b) still applies formally, except that $\Phi_\delta(z)$ are the eigenfunctions in the z domain arising now from the eigenvalue problem associated with g_z in the λ_u plane, while $\Phi_\alpha(u)$ are eigenfunctions in the u domain arising from the eigenvalue problem associated with g_u in the λ_v plane (in the latter, λ_u is held fixed at the eigenvalues arising from the eigenvalue problem in the z domain). As for (3.122b), the remarks concerning the form of the spectral representation apply here as well.

Alternative representations for Green's functions in spherical regions are constructed in a similar manner. On defining radial and angular characteristic Green's functions g_r, g_ϕ, and g_θ, one may rewrite the E mode Green's function in the following forms:

$$rr'G(\boldsymbol{r},\boldsymbol{r}') = \begin{cases} -\frac{1}{2\pi j}\sum_\beta \Phi_\beta(\phi)\Phi_\beta^*(\phi') \oint_{C_\theta} g_\theta(\theta,\theta';\beta^2;\lambda_\theta) g_r(r,r';\lambda_\theta)\, d\lambda_\theta, \\ \left(-\frac{1}{2\pi j}\right)^2 \oint_{C_\phi}\oint_{C_\theta} g_\phi(\phi,\phi';\lambda_\phi) g_\theta(\theta,\theta';\lambda_\phi;\lambda_\theta) g_r(r,r';\lambda_\theta)\, d\lambda_\phi\, d\lambda_\theta, \\ +\frac{1}{2\pi j}\sum_\beta \Phi_\beta(\phi)\Phi_\beta^*(\phi') \oint_{C_r} g_\theta(\theta,\theta';\beta^2;\lambda_\theta) g_r(r,r';\lambda_\theta)\, d\lambda_\theta, \\ \sum_\beta \Phi_\beta(\phi)\Phi_\beta^*(\phi') \sum_s R_s(r)\bar{R}_s(r') g_\theta(\theta,\theta';\beta^2;\lambda_s), \qquad \text{etc.} \end{cases} \tag{3.126}$$

The dependence of g_θ on the two parameters $\lambda_\phi = \beta^2$ and $\lambda_\theta = p(p+1)$ has been exhibited explicitly, and C_θ, C_r, and C_ϕ denote contours that enclose in the positive sense all of (and only) the singularities of g_θ, g_r, and g_ϕ in the complex λ_θ and λ_ϕ planes, respectively. The third equation of (3.126) follows from the first of (3.126) by contour deformation about the singularities of g_r, and the fourth of (3.126) results by evaluating the integral in terms of the radial eigenfunctions $R_s(r)$ and the adjoint functions $\bar{R}_s(r)$:

$$r'^2\delta(r-r') = \frac{1}{2\pi j}\oint_{C_r} g_r(r,r';\lambda)\, d\lambda = \sum_s R_s(r)\bar{R}_s(r'). \tag{3.127}$$

In addition to Section 1.5 of [14], detailed applications of the characteristic Green's function method for construction of alternative representations for $G(\boldsymbol{r},\boldsymbol{r}')$ may be found in Sections 5.6a, 5.7b, 6.7 and 6.8 of [14]. Directly analogous considerations can be applied to the scalar function $\mathscr{S}'$ defined in (2.3.24) or (2.3.39) of [14], in which case an additional pole singularity exists in the complex λ_u and (or) λ_v plane because of the presence of the $1/k_{ti}^2$ factor. Although the examples above involve primarily the electromagnetic E mode problem, construction of the electromagnetic H mode Green's functions proceeds similarly.

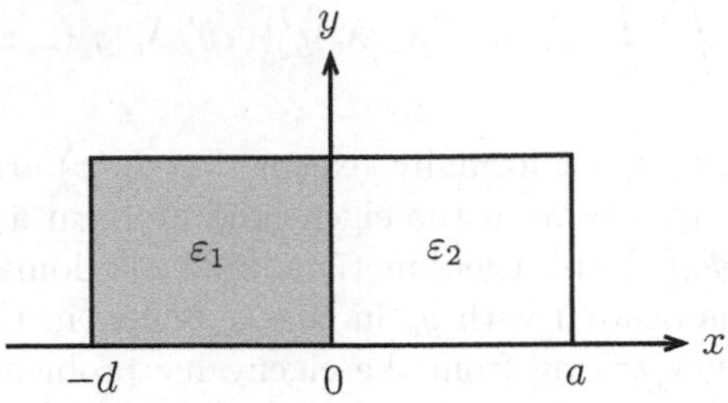

Fig. 3.6. Example of rectangular regions (bounded in x) partially filled with dielectric; a PEC is present at $x = -d$, $x = a$.

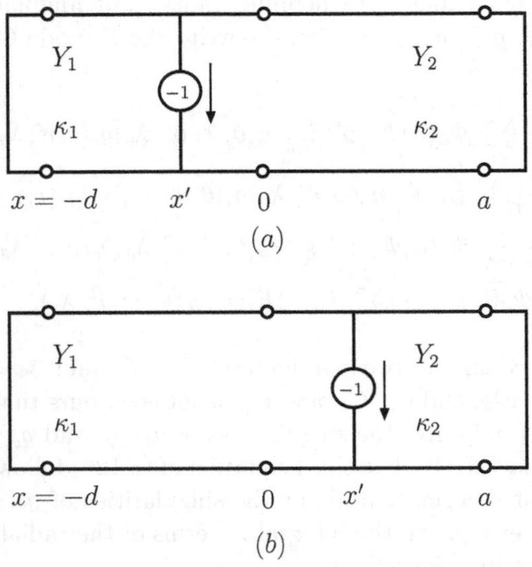

Fig. 3.7. Equivalent transmission line representations for TE modes along x.

XI 1D Characteristic Green's Function and Eigenfunction

The characteristic Green's function (GF) method for solving eigenvalue problems in closed and open regions is now applied to composite rectangular cross sections (for cylindrical and spherical cross sections see [14], Section 3.4c). We shall deal only with closed rectangular geometries in order to illustrate the procedure. For open regions characterized by unbounded x–domains extending to ∞, $-\infty$, or both, see [14].
We consider the composite cross sections shown in Figure 3.6, which are all characterized by the same one-dimensional eigenvalue problem in the x domain. The various media contain a piecewise constant lossless truncated dielectric

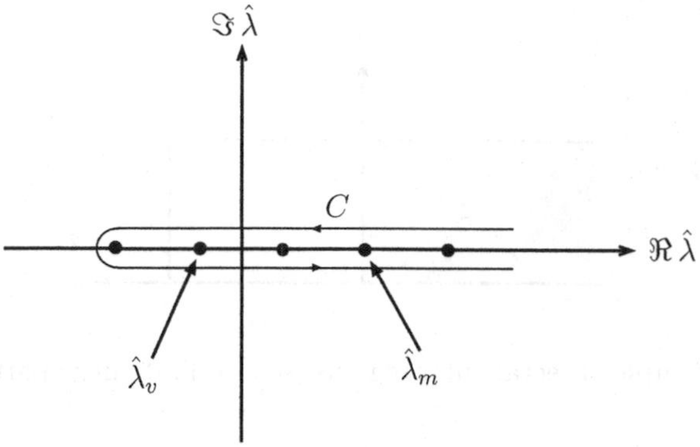

Fig. 3.8. Complex λ-plane singularities and integration contour.

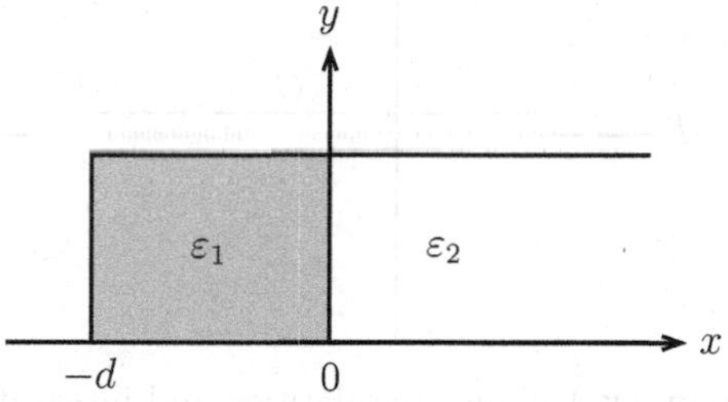

Fig. 3.9. Example of rectangular regions (semi-infinite in x) partially filled with dielectric.

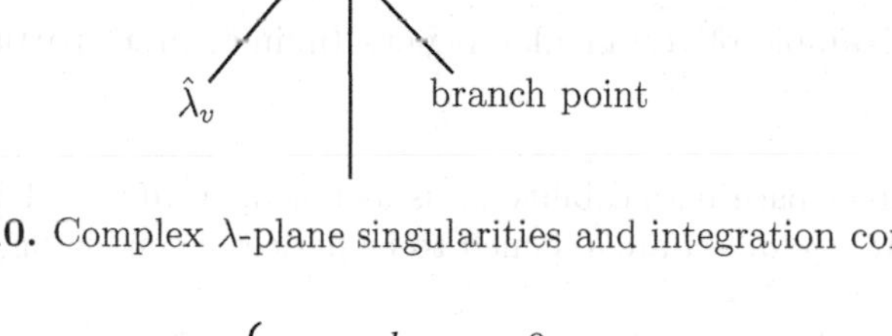

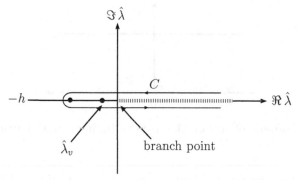

Fig. 3.10. Complex λ-plane singularities and integration contour.

$$\epsilon(x) = \begin{cases} \epsilon_1, & -d < x < 0 \\ \epsilon_2, & 0 < x < a \end{cases}, \quad \epsilon_1 > \epsilon_2 \tag{3.128}$$

which leads to a discontinuous representation of the eigenfunctions. The eigenvalue problems in the y domain are those appropriate to a homogeneous medium.

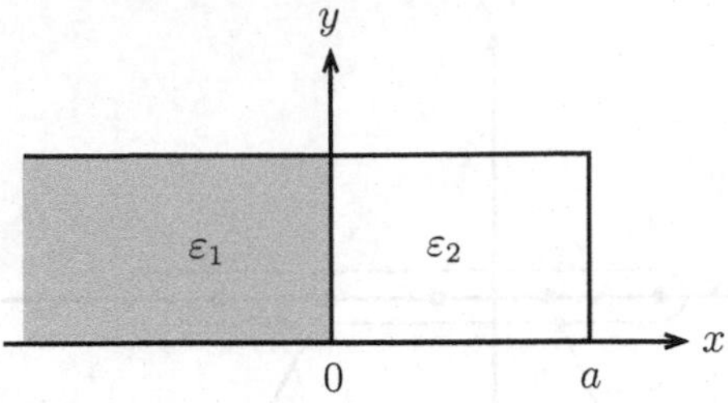

Fig. 3.11. Example of rectangular regions (semi-infinite in x) partially filled with dielectric.

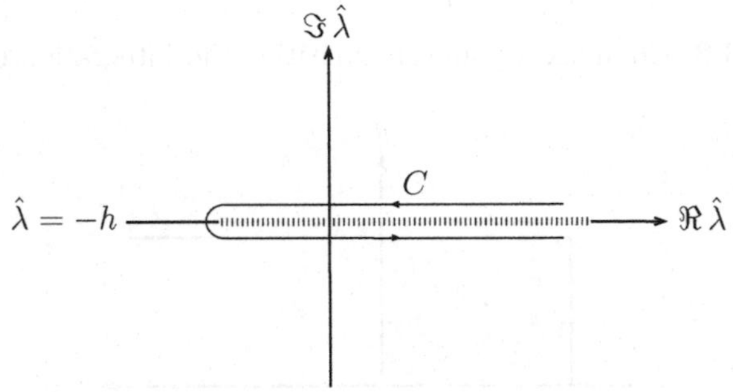

Fig. 3.12. Complex λ-plane singularities and integration contour.

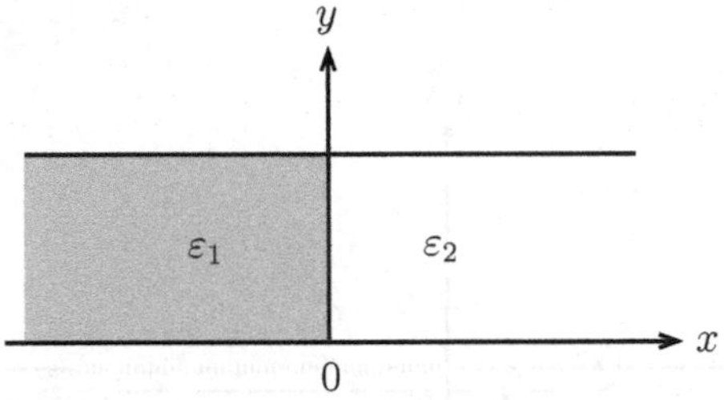

Fig. 3.13. Example of rectangular regions (infinite in x) partially filled with dielectric.

A constant, free-space permeability μ_0 is assumed, so $\bar{\mu}(x) = 1$ in (3.99b), and the surfaces at $x = a, -d$ are assumed to be perfectly conducting.

H Modes (in x)

The network configuration descriptive of the H mode characteristic GF problem is shown in Figure 3.7, where we distinguish between source locations in media 1 and 2, respectively. The relevant propagation constants and characteristic admittances are denoted, respectively, by $k_{x1} \equiv \kappa_1$, Y_1 and $k_{x2} = \kappa_2$, Y_2. From (3.100), with $\bar{\mu} = 1$, it is noted that the homogeneous equation defines standing-wave functions c and s,

$$\left[\frac{d^2}{dx^2} + \kappa^2(x,\lambda)\right] \left\{ \begin{matrix} c(x) \\ s(x) \end{matrix} \right\} = 0, \tag{3.129a}$$

where

$$\kappa^2(x,\lambda) = \begin{cases} \kappa_1^2(\lambda) = k_1^2 + \lambda, & -d < x < 0 \\ \kappa_2^2(\lambda) = k_2^2 + \lambda, & 0 < x < a \end{cases}, \quad k_{1,2}^2 = \omega^2 \mu_0 \epsilon_{1,2} > 0. \tag{3.129b}$$

The solutions are

$$\begin{aligned} c(x) &= \cos \kappa_1 x, \quad s(x) = \frac{1}{\kappa_1} \sin \kappa_1 x, \qquad -d < x < 0, \\ c(x) &= \cos \kappa_2 x, \quad s(x) = \frac{1}{\kappa_2} \sin \kappa_2 x, \qquad 0 < x < a. \end{aligned} \tag{3.130}$$

Since $\overleftarrow{Y}_T = \infty = \overrightarrow{Y}_T$ for the perfectly conducting terminations at $x = -d,\ a$, it follows from (3.106b) that

$$\omega\mu_0 \overrightarrow{Y}(0) = -j\kappa_2 \cot \kappa_2 a, \quad \omega\mu_0 \overleftarrow{Y}(0) = -j\kappa_1 \cot \kappa_1 d, \tag{3.131}$$

where $\kappa/\omega\mu_0$ is the H mode characteristic admittance. Thus, from (3.105)

$$\overrightarrow{V}(x) = \begin{cases} \overrightarrow{V}_2(x) = \dfrac{\sin \kappa_2(a-x)}{\sin \kappa_2 a}, & 0 < x < a, \\ \overrightarrow{V}_1(x) = \cos \kappa_1 x - \frac{\kappa_2}{\kappa_1} \cot \kappa_2 a \sin \kappa_1 x, & -d < x < 0, \end{cases} \tag{3.132a}$$

$$\overleftarrow{V}(x) = \begin{cases} \overleftarrow{V}_2(x) = \cos \kappa_2 x + \frac{\kappa_1}{\kappa_2} \cot \kappa_1 d \sin \kappa_2 x, & 0 < x < a, \\ \overleftarrow{V}_1(x) = \dfrac{\sin \kappa_1(x+d)}{\sin \kappa_1 d}, & -d < x < 0. \end{cases} \tag{3.132b}$$

For subsequent application it will be convenient to employ the traveling-wave formulation:

$$\overleftarrow{V}_2(x) = \frac{1}{1 + \overleftarrow{\Gamma}_2(0)} [e^{j\kappa_2 x} + \overleftarrow{\Gamma}_2(0) e^{-j\kappa_2 x}], \quad 0 < x < a, \tag{3.133}$$

where the reflection coefficient $\overleftarrow{\Gamma}_2(0)$ looking to the left at $x = +0$ is given by

$$\overleftarrow{\Gamma}_2(0) = \frac{Y_{02} - \overleftarrow{Y}(0)}{Y_{02} + \overleftarrow{Y}(0)} = \frac{\kappa_2 + j\kappa_1 \cot \kappa_1 d}{\kappa_2 - j\kappa_1 \cot \kappa_1 d}, \quad Y_{02} = \frac{\kappa_2}{\omega\mu_0}. \tag{3.134}$$

The H mode characteristic Green's function $g''(x, x'; \lambda)$ can now be written down directly from (3.106a). In view of the discontinuous representation of $\overleftrightarrow{V}(x)$ for

$x > 0$ and $x < 0$, g'' is represented discontinuously about $x = 0$. For a source location as in Figure 3.7(a),

$$g''(x, x'; \lambda) = \begin{cases} \dfrac{\overleftarrow{V}_1(x_<)\overrightarrow{V}_1(x_>)}{j\omega\mu_0 \overleftrightarrow{Y}(0)}, & -d < x < 0, \quad -d < x' < 0, \\ \dfrac{\overleftarrow{V}_1(x')\overrightarrow{V}_2(x)}{j\omega\mu_0 \overleftrightarrow{Y}(0)}, & 0 < x < a, \quad -d < x' < 0; \end{cases} \tag{3.135a}$$

whereas for the source location in Figure 3.7(b),

$$g''(x, x'; \lambda) = \begin{cases} \dfrac{\overleftarrow{V}_1(x)\overrightarrow{V}_2(x')}{j\omega\mu_0 \overleftrightarrow{Y}(0)}, & -d < x < 0, \quad 0 < x' < a, \\ \dfrac{\overleftarrow{V}_2(x_<)\overrightarrow{V}_2(x_>)}{j\omega\mu_0 \overleftrightarrow{Y}(0)}, & 0 < x < a, \quad 0 < x' < a. \end{cases} \tag{3.135b}$$

Equations (3.135) can be combined in the single formula

$$g''(x, x'; \lambda) = \frac{\overleftarrow{V}_\Omega(x_<)\overrightarrow{V}_\Omega(x_>)}{j\omega\mu_0 \overleftrightarrow{Y}(0)}, \quad \overleftrightarrow{Y}(0) = \overleftarrow{Y}(0) + \overrightarrow{Y}(0), \tag{3.136}$$

where the subscript Ω stands for 1 or 2 if the corresponding variable x or x' lies in the range $-d$ to 0 or 0 to a, respectively. To assure that the solution for g'' is unique, the restriction $\mathrm{i}\lambda \neq 0$ (i.e., $\mathrm{i}\kappa_1^2 \neq 0$, $\mathrm{i}\kappa_2^2 \neq 0$) is implied.

The singularities of g'' in the complex λ plane consist of real simple poles at the zeros of $\overleftrightarrow{Y}(0)$. Although g'' is a function of $\kappa_{1,2}$, and, from (3.129b),

$$\kappa_{1,2} = \sqrt{\lambda + k_{1,2}^2}, \tag{3.137}$$

no branch-point singularities exist at $\lambda = -k_{1,2}^2$, since $\overleftrightarrow{V}_\Omega$, $\overleftrightarrow{Y}(0)$ and therefore g'' are even functions of $\kappa_{1,2}$ [see (3.131)–(3.133)]. Thus, a power-series expansion about $\kappa_1 = 0$ or $\kappa_2 = 0$ comprises only integral powers of κ_1^2 or κ_2^2 and hence integral powers of λ, so the regularity of g'' in the neighborhood of the points $\lambda = -k_{1,2}^2$ is assured. From (3.131) the zeros λ_m of $\overleftrightarrow{Y}(0, \lambda)$ are specified implicitly by the transcendental equation

$$\kappa_2 \cot \kappa_2 a = -\kappa_1 \cot \kappa_1 d, \tag{3.138a}$$

$$\kappa_2^2 = \lambda + k_2^2 = \hat{\lambda}, \quad \kappa_1^2 = \lambda + k_1^2 = \hat{\lambda} + h, \quad h = k_1^2 - k_2^2 > 0. \tag{3.138b}$$

For real values of κ_1 and κ_2 (i.e., $\hat{\lambda} > 0$), (3.138a) has an infinite number of solutions to be denoted by κ_{1m}, κ_{2m} (only positive roots κ_{1m} and κ_{2m} need be considered since negative values leads to the same λ_m). For imaginary values of κ_1 and κ_2 ($\hat{\lambda} < -h$), (3.138a) becomes

$$|\kappa_2| \coth[|\kappa_2|a] = -|\kappa_1| \coth[|\kappa_1|d], \quad \kappa_1, \kappa_2 \text{ imaginary.} \tag{3.139a}$$

Since the left-hand side of (3.139a) is positive while the right-hand side is negative, no solution exists. However, for real κ_1 and imaginary κ_2 ($-h < \hat{\lambda} < 0$), (3.138a) can have roots $\kappa_{1\nu}$, $|\kappa_{2\nu}|$:

$$r_\nu \cot r_\nu = -t_\nu \coth\left(\frac{a}{d} t_\nu\right), \quad r_\nu^2 + t_\nu^2 = hd^2 = \left(1 - \frac{\varepsilon_2}{\varepsilon_1}\right)(k_1 d)^2, \tag{3.139b}$$

where

$$\kappa_{1\nu} \equiv r_\nu > 0; \quad |\kappa_{2\nu}|d \equiv t_\nu, \quad k_{2\nu} \text{ imaginary.} \tag{3.139c}$$

The spectral representation of the delta function is now obtained by integrating the characteristic Green's function g'' in (3.136) along the contour C shown in Figure 3.8 enclosing all singularities:

$$\delta(x - x') = -\frac{1}{2\pi j} \oint_C g''(x, x'; \lambda)\, d\lambda, \tag{3.140a}$$

$$= \sum_\nu \hat{\psi}_{\nu\Omega}(x)\hat{\psi}^*_{\nu\Omega}(x') + \sum_m \hat{\psi}_{m\Omega}(x)\hat{\psi}^*_{m\Omega}(x'), \quad -d < \begin{Bmatrix} x \\ x' \end{Bmatrix} < a, \tag{3.140b}$$

where the contributions for $-h < \hat{\lambda}_\nu < 0$ and $\hat{\lambda}_m > 0$ have been exhibited separately. From (3.107f) and (3.132) and (3.133) one obtains, for the orthonormal eigenfunctions $\hat{\psi}_{\nu\Omega}$ and $\hat{\psi}_{m\Omega}$,

$$\hat{\psi}_{\nu 1}(x) = \frac{1}{A_\nu} \frac{\sin[r_\nu(x/d + 1)]}{\sin r_\nu}, \quad 0 < r_\nu < \sqrt{h}d, \quad -d < x < 0, \tag{3.141a}$$

$$\hat{\psi}_{\nu 2}(x) = \frac{1}{A_\nu} \frac{\sinh[t_\nu \alpha(1 - x/a)]}{\sinh(t_\nu \alpha)}, \quad \alpha = \frac{a}{d}, \quad 0 < x < a, \tag{3.141b}$$

where

$$A_\nu^2 = \omega\mu_0 \frac{\partial}{\partial \lambda_\nu} \overleftrightarrow{B}(0, \lambda_\nu) = \frac{d}{2}\left[\frac{\coth(t_\nu \alpha)}{t_\nu r_\nu^2} hd^2 + \csc^2 r_\nu - \alpha \operatorname{csch}^2(t_\nu \alpha)\right]. \tag{3.141c}$$

Similarly,

$$\hat{\psi}_{m1}(x) = \frac{1}{A_m} \frac{\sin \kappa_{1m}(x + d)}{\sin \kappa_{1m} d}, \quad \kappa_{1m} > 0, \quad -d < x < 0, \tag{3.142a}$$

$$\hat{\psi}_{m2}(x) = \begin{cases} \dfrac{1}{A_m} \dfrac{\sin \kappa_{2m}(a - x)}{\sin \kappa_{2m} a} \\[2ex] \dfrac{1}{A_m \left[1 + \overleftarrow{\Gamma}_m(0)\right]} \left[e^{j\kappa_{1m} x} + \overleftarrow{\Gamma}_m(0)^{-j\kappa_{2m} x}\right], \end{cases} \quad \kappa_{2m} > 0, \quad 0 < x < a, \tag{3.142b}$$

with

$$A_m^2 = \omega\mu_0 \frac{\partial}{\partial\lambda_m} \overleftrightarrow{B}(0,\lambda_m) = \begin{cases} \dfrac{a}{2}\left[\left(1+\dfrac{1}{\alpha}\right) + \left(\dfrac{\kappa_{1m}^2}{\kappa_{2m}^2} + \dfrac{1}{\alpha}\right)\cot^2 \kappa_{1m} d \right. \\ \qquad\qquad \left. + \dfrac{h}{a\kappa_{1m}\kappa_{2m}^2} \cot \kappa_{1m} d\right], \\ \dfrac{d}{2}\left[(1+\alpha) + \left(\dfrac{\kappa_{2m}^2}{\kappa_{1m}^2} + \alpha\right)\cot^2 \kappa_{2m} a \right. \\ \qquad\qquad \left. - \dfrac{h}{d\kappa_{2m}\kappa_{1m}^2} \cot \kappa_{2m} a\right]. \end{cases} \tag{3.142c}$$

Equations (3.141) and (3.142) reduce to the special case of a homogeneously filled waveguide when (a) $h = 0 (\varepsilon_1 = \varepsilon_2)$, (b) $d = 0$, or (c) $a = 0$. Attention should be called to the different behavior of the eigenfunctions $\hat{\psi}_\nu(x)$ in (3.141) and $\hat{\psi}_m(x)$ in (3.142). While $\hat{\psi}_m(x)$ is represented by an oscillating function over the entire region $-d < x < a$, $\hat{\psi}_\nu(x)$ behaves in this manner only in the dielectric ε_1, (note: $\varepsilon_1 > \varepsilon_2$). In the remaining interval $0 < x < a$, $\hat{\psi}_\nu$ decays away from the interface $x = 0$. Viewed in modal terms with respect to propagation along z, the fields corresponding to the $\hat{\psi}_\nu$ are essentially confined within the dielectric slab while the fields derived from the $\hat{\psi}_m$ fill the entire waveguide cross section. The former are termed "trapped" modes and their existence depends entirely on the presence of the dielectric; the latter may be regarded as perturbations about the dielectric-free case.

E modes (in x).

The solution for the E mode characteristic Green's function $g'(x,x';\lambda)$ and the associated orthonormal eigenfunctions is similar to the above except for duality replacements [see (3.108)–(3.113)]. The results are summarized below.

Characteristic Green's function

$$g'(x,x';\lambda) = \frac{\overleftarrow{I}_\Omega(x_<)\overrightarrow{I}_\Omega(x_>)}{j\omega\varepsilon_0 \overleftrightarrow{Z}(0)}, \quad \overleftrightarrow{Z}(0) = \overleftarrow{Z}(0) + \overrightarrow{Z}(0). \tag{3.143a}$$

Standing-Wave Functions

$$\begin{aligned} c(x) &= \cos\kappa_1 x, \quad s(x) = \frac{\bar{\varepsilon}_1}{\kappa_1}\sin\kappa_1 x, \quad -d < x < 0, \\ c(x) &= \cos\kappa_2 x, \quad s(x) = \frac{\bar{\varepsilon}_2}{\kappa_2}\sin\kappa_2 x, \quad 0 < x < a, \end{aligned} \tag{3.143b}$$

$$\omega\varepsilon_0 \overrightarrow{Z}(0) = j\frac{\kappa_2}{\bar{\varepsilon}_2}\tan\kappa_2 a, \quad \omega\varepsilon_0 \overleftarrow{Z}(0) = j\frac{\kappa_1}{\bar{\varepsilon}_1}\tan\kappa_1 d, \quad \bar{\varepsilon}_{1,2} = \frac{\varepsilon_{1,2}}{\varepsilon_0} \tag{3.143c}$$

$$\kappa_1^2 = k_1^2 + \lambda, \quad \kappa_2^2 = k_2^2 + \lambda. \tag{3.143d}$$

$$\left.\begin{aligned}\overrightarrow{I}_1(x) &= \cos\kappa_1 x + \frac{\bar{\varepsilon}_1\kappa_2}{\bar{\varepsilon}_2\kappa_1}\tan\kappa_2 a\sin\kappa_1 x\\ \overleftarrow{I}_1(x) &= \frac{\cos\kappa_1(x+d)}{\cos\kappa_1 d}\end{aligned}\right\} -d < x < 0, \tag{3.143e}$$

$$\left.\begin{aligned}\overrightarrow{I}_2(x) &= \frac{\cos\kappa_2(a-x)}{\cos\kappa_2 a}\\ \overleftarrow{I}_2(x) &= \cos\kappa_2 x - \frac{\bar{\varepsilon}_2\kappa_1}{\bar{\varepsilon}_1\kappa_2}\tan\kappa_1 d\sin\kappa_2 x\end{aligned}\right\} 0 < x < a. \tag{3.143f}$$

Singularities of g': Simple real poles at

$$\frac{\bar{\varepsilon}_1}{\bar{\varepsilon}_2}\kappa_{2m}\tan\kappa_{2m}a = -\kappa_{1m}\tan\kappa_{1m}d, \quad \kappa_{1m},\kappa_{2m} > 0, \tag{3.144a}$$

and at

$$\begin{gathered}\frac{\bar{\varepsilon}_2}{\bar{\varepsilon}_1}r_\nu\tan r_\nu = t_\nu\tanh(t_\nu\alpha), \quad r_\nu^2 + t_\nu^2 = hd^2, \quad \alpha = \frac{a}{d}, h = k_1^2 - k_2^2,\\ \kappa_{1\nu}d \equiv r_\nu > 0; \quad |\kappa_{2\nu}|d \equiv t_\nu, \ \kappa_{2\nu} \text{ imaginary.}\end{gathered} \tag{3.144b}$$

Equation (3.144a) has an infinite number of solutions and (3.144b) has a finite number. The low-frequency cutoff found for the H mode solutions $\hat{\psi}_\nu$ is absent in the E mode case.

Delta-function representation $\left(d < \begin{Bmatrix} x \\ x' \end{Bmatrix} < a\right)$,

$$\begin{aligned}\bar{\varepsilon}(x')\delta(x-x') &= -\frac{1}{2\pi j}\oint_C g'(x,x';\lambda)\,d\lambda\\ &= \sum_\nu \hat{\Phi}_{\nu\Omega}(x)\hat{\Phi}^*_{\nu\Omega}(x') + \sum_m \hat{\Phi}_{m\Omega}(x)\hat{\Phi}^*_{m\Omega}(x'),\end{aligned} \tag{3.145a}$$

with the subscript Ω defined as under (3.136), and

$$\hat{\Phi}_{\nu 1}(x) = \frac{\cos[r_\nu(x/d+1)]}{A_\nu\cos r_\nu}, \quad 0 < r_\nu < \sqrt{h}d, \quad -d < x < 0, \tag{3.145b}$$

$$\hat{\Phi}_{\nu 2}(x) = \frac{\cosh[t_\nu\alpha(1-x/a)]}{A_\nu\cosh t_\nu\alpha}, \quad 0 < x < a, \tag{3.145c}$$

$$A_\nu^2 = \omega\varepsilon_0\frac{\partial}{\partial\lambda_\nu}\overleftrightarrow{X}(0,\lambda_\nu) = \frac{d}{2}\left[\frac{\tanh(t_\nu\alpha)}{r_\nu^2 t_\nu\bar{\varepsilon}_2}hd^2 + \frac{\sec^2 r_\nu}{\bar{\varepsilon}_1} + \frac{\alpha}{\bar{\varepsilon}_2}\operatorname{sech}^2(t_\nu\alpha)\right], \tag{3.145d}$$

while

$$\hat{\Phi}_{m1}(x) = \frac{\cos\kappa_{1m}(x+d)}{A_m \cos\kappa_{1m}d}, \quad \kappa_{1m} > 0, \quad -d < x < 0, \tag{3.145e}$$

$$\hat{\Phi}_{m2}(x) = \frac{\cos\kappa_{2m}(a-x)}{A_m \cos\kappa_{2m}a}, \quad \kappa_{2m} > 0, \quad 0 < x < a, \tag{3.145f}$$

$$A_m^2 = \frac{a}{2\bar{\varepsilon}_1}\left[\frac{\bar{\varepsilon}_1}{\bar{\varepsilon}_2} + \frac{1}{\alpha} + \left(\frac{\kappa_{1m}^2\bar{\varepsilon}_2}{\kappa_{2m}^2\bar{\varepsilon}_1} + \frac{1}{\alpha}\right)\tan^2\kappa_{1m}d - \frac{h}{a\kappa_{1m}\kappa_{2m}^2}\tan\kappa_{1m}d\right]. \tag{3.145g}$$

The physical distinction between the mode fields corresponding to the $\hat{\Phi}_\nu$ and $\hat{\Phi}_m$ is the same as discussed in connection with the H modes.

Employing (3.145a), one may represent a suitable function $F(x)$ in the interval $-d < x < a$ as follows:

$$\begin{aligned} F(x) &= \int_{-d}^{a} \frac{F(x')}{\bar{\varepsilon}(x')}\bar{\varepsilon}(x')\delta(x-x')\,dx' \\ &= \begin{cases} \sum_\nu f_\nu \hat{\Phi}_{\nu 1}(x) + \sum_m f_m \hat{\Phi}_{m1}(x), & -d < x \le 0, \\ \sum_\nu f_\nu \hat{\Phi}_{\nu 2}(x) + \sum_m f_m \hat{\Phi}_{m2}(x), & 0 \le x < a, \end{cases} \end{aligned} \tag{3.146a}$$

where

$$f_\nu = \frac{1}{\bar{\varepsilon}_1}\int_{-d}^{0} F(x')\hat{\Phi}_{\nu 1}^*(x')\,dx' + \frac{1}{\bar{\varepsilon}_2}\int_0^a F(x')\hat{\Phi}_{\nu 2}^*(x')\,dx', \tag{3.146b}$$

$$f_m = \frac{1}{\bar{\varepsilon}_1}\int_{-d}^{0} F(x')\hat{\Phi}_{m1}^*(x')\,dx' + \frac{1}{\bar{\varepsilon}_2}\int_0^a F(x')\hat{\Phi}_{m2}^*(x')\,dx', \tag{3.146c}$$

and the asterisk denotes the complex conjugate.

Semiinfinite x domain

As $a \to \infty$ in Figure 3.6, one obtains the open cross-section configurations in Figure 3.9. The eigenfunctions appropriate to this case can be obtained as a limiting case of those for finite a.

H modes (in x) ($a \to \infty$). As $a \to \infty$, the resonances κ_{1m} and κ_{2m} in (3.138a), with $\kappa_{2m} > 0$, coalesce into a continuous spectrum, while those in (3.139b) remain discrete and satisfy the equation

$$r_\nu \cot r_\nu = -t_\nu, \quad r_\nu^2 + t_\nu^2 = hd^2, \quad \text{as } a \to \infty, \tag{3.147a}$$

Moreover, from (3.141c),

$$A_\nu^2 \to \frac{d}{2}\frac{hd^2}{r_\nu^2}\left(1 + \frac{1}{t_\nu}\right), \tag{3.147b}$$

while, from (3.142c),

$$A_m^2 \to A_\xi^2 = \frac{a}{2}\left(1 + \frac{\xi_1^2}{\xi^2}\cot^2 \xi_1 d\right) = \frac{2a}{[q + \overleftarrow{\Gamma}_2(\xi,0)][1 + \overleftarrow{\Gamma}_2(\xi,0)^*]}, \tag{3.147c}$$

where $\overleftarrow{\Gamma}_2(\xi,0)$ is given in (3.133). In the last equation the continuous variables ξ_1 and ξ have been defined as the limiting values of κ_{1m} and κ_{2m} as $a \to \infty$:

$$\kappa_{2m} \to \xi, \quad \kappa_{1m} \to \xi_1 = \sqrt{\xi^2 + h}, \quad 0 < \xi < \infty, \quad a \to \infty. \tag{3.147d}$$

Upon noting that the increment between successive resonances, $\Delta\xi_m = \xi_{m+1} - \xi_m \to \pi/a$ as $a \to \infty$, the continuous limit in (3.140b) yields

$$\delta(x - x') = \sum_\nu \hat{\psi}_{\nu\Omega}(x)\hat{\psi}^*_{\nu\Omega}(x') + \int_0^\infty \hat{\psi}_\Omega(\xi,x)\hat{\psi}^*_\Omega(\xi,x')\,d\xi,$$
$$-d < \begin{Bmatrix} x \\ x' \end{Bmatrix} < \infty, \quad \Omega = 1,2, \tag{3.148a}$$

where, in view of (3.141), (3.142), and (3.147), one has, for the discrete spectrum,

$$\hat{\psi}_{\nu1}(x) = \frac{1}{A_\nu}\frac{\sin[r_\nu((x/d)+1)]}{\sin r_\nu}, \quad 0 < r_\nu < \sqrt{h}d, \quad -d < x < 0, \tag{3.148b}$$

$$\hat{\psi}_{\nu2}(x) = \frac{1}{A_\nu}e^{-t_\nu x/d}, \quad 0 < x < \infty. \tag{3.148c}$$

As in (3.136), $\Omega = 1$ for x or x' between $-d$ and 0, while $\Omega = 2$ for x or x' between 0 and ∞. Just as in the closed region, the magnitude of $\hat{\psi}_{\nu1}$ oscillates while that of $\hat{\psi}_{\nu2}$ decreases exponentially for $x > 0$. Thus, the field of such a mode is confined again to the region $-d < x < 0$ occupied by the dielectric ε_1. Modes traveling in the z direction with this transverse field behavior are characterized as "trapped waves", or "surface waves", since the field appears to be trapped inside the dielectric with the larger permittivity and guided by the dielectric surface. For the continuous spectrum,

$$\hat{\psi}_1(\xi,x) = \frac{\sin \xi_1(x+d)}{\sqrt{2\pi}\sin \xi_1 d}[1 + \overleftarrow{\Gamma}_2(0,\xi)], \quad 0 < \xi < \infty, \quad -d < x < 0, \tag{3.148d}$$

$$\hat{\psi}_2(\xi,x) = \frac{1}{\sqrt{2\pi}}[e^{j\xi x} + \overleftarrow{\Gamma}_2(0,\xi)e^{-j\xi x}], \quad 0 < x < \infty, \tag{3.148e}$$

where

$$\overleftarrow{\Gamma}_2(0,\xi) = \frac{\xi + j\xi_1 \cot \xi_1 d}{\xi - j\xi_1 \cot \xi_1 d}, \quad \xi_1^2 = h + \xi^2 = (k_1^2 - k_2^2) + \xi^2. \tag{3.148f}$$

The traveling–wave representation for $\hat{\psi}_2$ derived as a limiting case of (3.142b), has a significant physical interpretation. For the assumed time dependence $\exp(+j\omega t)$, the contribution from the first term inside the brackets in (3.148e) constitutes a properly normalized (incident) free-space plane-wave mode traveling in the $-x$ direction, while the second term comprises the wave reflected at $x = 0$ with reflection coefficient $\overleftarrow{\Gamma}_2(0, \xi)$. Thus, the continuous spectrum for $x > 0$ is obtained by adding to a properly normalized incident wave a reflected wave so adjusted that the boundary conditions at $x = 0$ are satisfied.

The delta-function representation in (3.148a) could also have been deduced directly from the characteristic Green's function. As a $a \to \infty$ and since $\mathrm{i}\kappa_2 \neq 0$, the standing wave in (3.132) goes over into a traveling wave. In this transition, the restriction $\mathrm{i}\kappa_2 < 0$ appropriate to the assumed time dependence $\exp(+j\omega t)$ must be observed and yields the following (bounded) result for $x > 0$:

$$\overrightarrow{V}_2(x) \to e^{-j\kappa_2 x}, \quad \kappa_2 = \sqrt{k_2^2 + \lambda} = \sqrt{\hat{\lambda}}, \quad \mathrm{i}\kappa_2 < 0, \tag{3.149a}$$

and

$$\overrightarrow{V}_1(x) \to \cos\kappa_1 x - j\frac{\kappa_2}{\kappa_1}\sin\kappa_1 x, \quad \kappa_1 = \sqrt{\hat{\lambda} + h}, \quad h = k_1^2 - k_2^2. \tag{3.149b}$$

Moreover, from (3.131),

$$\overrightarrow{Y}(0) \to \frac{\kappa_2}{\omega\mu_0}, \quad \text{i.e.,}\ j\omega\mu_0\overleftrightarrow{Y}(0) = j\kappa_2 + \kappa_1\cot\kappa_1 d. \tag{3.149c}$$

The $\overleftarrow{V}_{1,2}(x)$ are still given by (3.132b). $\overleftrightarrow{V}_\Omega$ and $\overleftrightarrow{Y}(0)$ remain even functions of κ_1 but not of κ_2. $\lambda = -k_1^2$ is therefore a regular point in the complex λ-plane. On the other hand, an expansion of $g''(x, x'; \lambda)$ about the point $\lambda = -k_2^2$ contains integral powers of κ_2, so $\lambda + k_2^2 = \hat{\lambda} = 0$ is a branch point of order 1. If we define

$$\hat{\lambda} = |\hat{\lambda}|e^{j\gamma}, \quad \sqrt{\hat{\lambda}} = |\sqrt{\hat{\lambda}}|e^{j\gamma/2}, \tag{3.150}$$

the convergence requirement $\mathrm{i}\sqrt{\hat{\lambda}} < 0$ in (3.149a) restricts the argument γ to the range $0 > \gamma > -2\pi$. To impose this condition on the entire top sheet, the spectral sheet, of the two-sheeted complex λ plane, one chooses a branch-cut along the positive real axis as shown in Figure 3.10.

The Green's function g'' may also have relevant pole singularities at the zeros of $\overleftrightarrow{Y}(0)$, namely when

$$j\kappa_2 = -\kappa_1\cot\kappa_1 d. \tag{3.151}$$

Solutions of (3.151) exist only for real values of κ_1 and imaginary values of $\kappa_2 = -j|\kappa_2|$ (i.e., $0 > \hat{\lambda} > -h$), leading to the transcendental equation (3.147a). The location of possible pole singularities is shown in Figure 3.10. Upon performing an integration as in (3.140a) about the contour C in Figure 3.10 enclosing all the singularities of g'' in the complex λ plane, one obtains after residue evaluation at the poles λ_ν the series in (3.148a), with $g''(x,x';\lambda)$ given by (3.136) and subject to the modifications in (3.149). The remaining contour integral about the branch cut can be written as

$$I = -\frac{1}{2\pi j}\int_{\infty e^{-j2\pi}}^{0} g''(x,x';\lambda)\,d\hat{\lambda} - \frac{1}{2\pi j}\int_{0}^{\infty e^{-j0}} g''(x,x';\lambda)\,d\hat{\lambda} \tag{3.152a}$$

$$= -\frac{1}{2\pi j}\int_{0}^{\infty e^{-j0}} [g''(x,x';\hat{\lambda}-k_2^2) - g''(x,x';\hat{\lambda}e^{-j2\pi}-k_2^2)]\,d\hat{\lambda}$$

$$= -\frac{1}{\pi}\mathrm{i}\int_{0}^{\infty e^{-j0}} g''(x,x';\hat{\lambda}-k_2^2)\,d\hat{\lambda}$$

$$= -\frac{2}{\pi}\mathrm{i}\int_{0}^{\infty} \xi g''(x,x';\xi^2-k_2^2)\,d\xi, \quad \xi^2=\hat{\lambda}. \tag{3.152b}$$

The transition from (3.152a) to (3.152b) is based on the property

$$g''(x,x';\hat{\lambda}e^{-j2\pi}) = g''(x,x';\hat{\lambda}^*) = g''^*(x,x';\hat{\lambda}), \quad \hat{\lambda} = |\hat{\lambda}|e^{-j0}, \tag{3.152c}$$

satisfied by g''. Upon substituting the appropriate representations for g'' into (3.152b), one obtains directly the continuous spectrum as in (3.148a).

E modes (in x) ($a \to \infty$)

The results for the E mode problem, obtained in direct analogy to those above, are summarized below:

$$\bar{\varepsilon}(x')\delta(x-x') = -\frac{1}{2\pi j}\oint_C g'(x,x';\lambda)\,d\lambda \tag{3.153a}$$

$$= \sum_\nu \hat{\Phi}_{\nu\Omega}(x)\hat{\Phi}^*_{\nu\Omega}(x') + \int_0^\infty \hat{\Phi}_\Omega(\xi,x)\hat{\Phi}^*_\Omega(\xi,x')\,d\xi, \tag{3.153b}$$

$$-d < \begin{matrix} x \\ x' \end{matrix} < \infty, \quad \Omega = 1,2,$$

where, for the discrete spectrum [see (3.148a) for definition of domains corresponding to $\Omega = 1,2$],

$$\hat{\Phi}_{\nu 1}(x) = \frac{\cos[r_\nu((x/d)+1)]}{A_\nu \cos r_\nu}, \quad 0 < r_\nu < \sqrt{h}d, \quad -d < x < 0, \tag{3.154a}$$

$$\hat{\Phi}_{\nu 2}(x) = \frac{e^{-t_\nu x/d}}{A_\nu}, \quad 0 < x < \infty. \tag{3.154b}$$

$$A_\nu^2 = \frac{d}{2}\left\{\left[1+\left(\frac{t_\nu}{r_\nu}\right)^2\right]\frac{1}{t_\nu\bar{\varepsilon}_2} + \left[1+\left(\frac{t_\nu\bar{\varepsilon}_1}{r_\nu\bar{\varepsilon}_2}\right)^2\right]\frac{1}{\bar{\varepsilon}_1}\right\}. \tag{3.154c}$$

Also, r_ν and t_ν are the solutions of the transcendental equations

$$\frac{\bar{\varepsilon}_2}{\bar{\varepsilon}_1} r_\nu \tan r_\nu = t_\nu, \qquad r_\nu^2 + t_\nu^2 = hd^2. \tag{3.154d}$$

The continuous spectrum is given by

$$\hat{\Phi}_1(\xi, x) = \sqrt{\frac{\bar{\varepsilon}_2}{2\pi}} \frac{\cos \xi_1 (x+d)}{\cos \xi_1 d} [1 - \overleftarrow{\Gamma}_2(0,\xi)], \quad \xi_1^2 = h + \xi^2,$$
$$0 < \xi < \infty, \quad -d < x < 0, \tag{3.155a}$$

$$\hat{\Phi}_2(\xi, x) = \sqrt{\frac{\bar{\varepsilon}_2}{2\pi}} [e^{j\xi x} - \overleftarrow{\Gamma}_2(0,\xi) e^{-j\xi x}], \quad 0 < x < \infty, \tag{3.155b}$$

$$\overleftarrow{\Gamma}_2(0,\xi) = \frac{\overleftarrow{Z}(0) - Z_{02}}{\overleftarrow{Z}(0) + Z_{02}} = \frac{j\xi_1 \tan \xi_1 d - \xi(\bar{\varepsilon}_1/\bar{\varepsilon}_2)}{j\xi_1 \tan \xi_1 d + \xi(\bar{\varepsilon}_1/\bar{\varepsilon}_2)}. \tag{3.155c}$$

If $d \to \infty$ in Figure 3.6, one obtains the semi-infinite configurations shown in Figure 3.11, which differ from those in Figure 3.9 in that the medium with the larger dielectric constant (ε_1) extends to infinity in the x direction.

H modes (in x) ($d \to \infty$)

As $d \to \infty$ in (3.138a), the resonances $\kappa_{1m}, \kappa_{2m} > 0$, coalesce into a continuous spectrum and the second series in the delta-function representation (3.140b) transforms into an integral analogous to that in (3.148a). However, in distinction to the case $a \to \infty$, the resonance parameters $\kappa_{1\nu}$, and $|\kappa_{2\nu}|$ in (3.139b) become continuous as $d \to \infty$. In tracing out the transition $d \to \infty$, one employs instead of (3.132a) the traveling-wave formulation similar to that in (3.133):

$$\overrightarrow{V}_1(x) = \frac{1}{1 + \overrightarrow{\Gamma}_1(0)} [e^{-j\kappa_1 x} + \overrightarrow{\Gamma}_1(0) e^{j\kappa_1 x}], \quad -d < x < 0, \tag{3.156a}$$

where the reflection coefficient $\overrightarrow{\Gamma}_1(0)$ seen to the right at $x = 0$ is given by

$$\overrightarrow{\Gamma}_1(0) = \frac{\kappa_1 + j\kappa_2 \cot \kappa_2 a}{\kappa_1 - j\kappa_2 \cot \kappa_2 a}. \tag{3.156b}$$

Since from (3.141c) and (3.142c),

$$A_m^2 \to A_{\xi_1}^2 = \frac{2d}{[1 + \overrightarrow{\Gamma}_1(\xi_1, 0)][1 + \overrightarrow{\Gamma}_1(\xi_1, 0)^*]}, \quad d \to \infty, \quad \sqrt{h} < \xi_1 < \infty, \tag{3.157a}$$

with

$$\overrightarrow{\Gamma}_1(\xi_1, 0) = \frac{\xi_1 + j\xi \cot \xi a}{\xi_1 - j\xi \cot \xi a}, \quad \xi = \sqrt{\xi_1^2 - h}, \quad h = k_1^2 - k_2^2 > 0, \tag{3.157b}$$

and

$$A_\nu^2 \to A_{\xi_1}^2, \quad d \to \infty, \quad 0 < \xi_1 < \sqrt{h}, \tag{3.157c}$$

one obtains via (3.140)–(3.142) and (3.156a) the delta-function representation:

$$\delta(x - x') = \int_0^\infty \hat{\psi}_\Omega(\xi_1, x)\hat{\psi}_\Omega^*(\xi_1, x')\, d\xi_1, \quad -\infty < \begin{Bmatrix} x \\ x' \end{Bmatrix} < a, \quad \Omega = 1, 2, \tag{3.158a}$$

where $-\infty < (x \text{ or } x') < 0$ for $\Omega = 1$ and $0 < (x \text{ or } x') < a$ for $\Omega = 2$, with

$$\hat{\psi}_1(\xi_1, x) = \frac{1}{\sqrt{2\pi}}[e^{-j\xi_1 x} + \overrightarrow{\Gamma}_1(\xi_1, 0)e^{j\xi_1 x}], \quad 0 < \xi_1 < \infty, \quad -\infty < x < a, \tag{3.158b}$$

$$\hat{\psi}_2(\xi_1, x) = \frac{\sin \xi(a - x)}{\sqrt{2\pi} \sin \xi a}[1 + \overrightarrow{\Gamma}_1(\xi_1, 0)], \quad 0 < x < a. \tag{3.158c}$$

It is noted that ξ is imaginary for $0 < \xi_1 < \sqrt{h}$.

To deduce (3.158a) directly from a characteristic Green's function analysis, one notes from (3.132) that as $d \to \infty$, with $\mathrm{i}\kappa_1 < 0$ appropriate to an $\exp(j\omega t)$ time dependence,

$$\overleftarrow{V}_1(x) \to e^{j\kappa_1 x}, \quad -\infty < x < 0, \tag{3.159a}$$

$$\overleftarrow{V}_2(x) \to \cos \kappa_2 x + j\frac{\kappa_1}{\kappa_2} \sin \kappa_2 x, \quad 0 < x < a, \tag{3.159b}$$

$$\omega\mu_0 \overleftarrow{Y}(0) \to \kappa_1. \tag{3.159c}$$

Since $g''(x, x'; \lambda)$, by (3.136) and (3.159), is an even function of κ_2 but not of κ_1, a branch-point singularity exists at $\kappa_1 = 0$ (i.e., $\lambda = -k_1^2$) in the complex λ plane. In analogy to (3.150), the restriction on the argument of λ on the spectral sheet is

$$\mathrm{i}\sqrt{\hat{\lambda} + h} < 0, \quad \text{i.e.,} -2\pi < \arg(\hat{\lambda} + h) < 0, \quad \hat{\lambda} = \lambda + k_2^2 = \kappa_2^2, \tag{3.160}$$

so that the branch cut is drawn from $\hat{\lambda} = -h$ to ∞ along the positive real axis in the $\hat{\lambda}$ plane (see Figure 3.12). To determine possible pole singularities we examine the resonance condition

$$j\omega\mu_0 \overleftrightarrow{Y}(0) = 0 = j\kappa_1 + \kappa_2 \cot \kappa_2 a. \tag{3.161}$$

Since (3.161) has no real solution λ_ν on the branch $\mathrm{i}\kappa_1 < 0$[13], no pole singularities exist, and the contour of integration is that shown in Figure 3.12. Thus, in analogy with (3.152),

[13] The corresponding discrete eigenfunctions, if they exist, must be square integrable (i.e., vanish at $x \to -\infty$). Since the problem is non-dissipative, any discrete eigenvalues must be real.

$$\delta(x-x') = -\frac{1}{2\pi j}\oint_C g''(x,x';\lambda)\,d\lambda \tag{3.162a}$$

$$= -\frac{2}{\pi}\mathrm{i}\int_0^\infty d\xi_1\xi_1 g''(x,x';\xi_1^2-k_1^2), \quad -\infty < \left\{\begin{matrix} x \\ x' \end{matrix}\right\} < a, \tag{3.162b}$$

which, upon insertion of g'' from (3.136), (3.132), and (3.159), yields (3.158a).
E modes (in x) ($d \to \infty$)
Spectral representation of delta function:

$$\bar{\varepsilon}(x')\delta(x-x') = -\frac{1}{2\pi j}g'(x,x';\lambda)\,d\lambda, \tag{3.163a}$$

$$= \int_0^\infty \hat{\Phi}_\Omega(\xi_1,x)\hat{\Phi}^*_\Omega(\xi_1,x')\,d\xi_1, \quad \Omega = 1,2 \tag{3.163b}$$

where $\Omega = 1$ when $-\infty < (x \text{ or } x') < 0$ while $\Omega = 2$ when $0 < (x \text{ or } x') < a$. The contour C in the complex λ plane is as shown in Figure 3.12, and from (3.145a) as $d \to \infty$,

$$\hat{\Phi}_1(\xi_1,x) = \sqrt{\frac{\bar{\varepsilon}_1}{2\pi}}[e^{-j\xi_1 x} - \overrightarrow{\Gamma}_1(\xi_1,0)e^{j\xi_1 x}], \quad 0 < \xi_1 < \infty, \quad -\infty < x < 0, \tag{3.164a}$$

$$\hat{\Phi}_2(\xi_1,x) = \sqrt{\frac{\bar{\varepsilon}_1}{2\pi}}[1 - \overrightarrow{\Gamma}_1(\xi_1,0)]\frac{\cos\xi(a-x)}{\cos\xi a}, \quad 0 < x < a, \tag{3.164b}$$

$$\overrightarrow{\Gamma}_1(\xi_1,0) = \frac{j\xi\tan\xi a - \xi_1(\bar{\varepsilon}_2/\bar{\varepsilon}_1)}{j\xi\tan\xi a + \xi_1(\bar{\varepsilon}_2/\bar{\varepsilon}_1)}, \quad \xi = \sqrt{\xi_1^2 - h}, \quad h = k_1^2 - k_2^2 > 0. \tag{3.164c}$$

Infinite x domain
Configurations comprising two dielectrics, semi-infinite in x, are shown in Figure 3.13
H modes (in x)
The characteristic Green's function for this case is given

$$g''(x,x';\lambda) = \frac{\overleftarrow{V}_\Omega(x_<)\overrightarrow{V}_\Omega(x_>)}{j\omega\mu_0\overleftrightarrow{Y}(0)}, \tag{3.165a}$$

with

$$\overleftarrow{V}_1(x) = e^{j\kappa_1 x}, \quad \kappa_1^2 = k_1^2 + \lambda, \quad \mathrm{i}\kappa_1 < 0, \tag{3.165b}$$

$$\overrightarrow{V}_2(x) = e^{-j\kappa_2 x}, \quad \kappa_2^2 = k_2^2 + \lambda, \quad \mathrm{i}\kappa_2 < 0, \tag{3.165c}$$

$$\overrightarrow{V}_1(x) = \frac{1}{1+\overrightarrow{\Gamma}_1(0)}\left[e^{-j\kappa_1 x} + \overrightarrow{\Gamma}_1(0)e^{j\kappa_1 x}\right], \quad \overrightarrow{\Gamma}_1(0) = \frac{\kappa_1 - \kappa_2}{\kappa_1 + \kappa_2}, \tag{3.166a}$$

$$\overleftarrow{V}_2(x) = \frac{1}{1+\overleftarrow{\Gamma}_2(0)}\left[e^{j\kappa_2 x} + \overleftarrow{\Gamma}_2(0)e^{-j\kappa_2 x}\right], \quad \overleftarrow{\Gamma}_2(0) = -\overrightarrow{\Gamma}_1(0), \tag{3.166b}$$

$$j\omega\mu_0 \overleftrightarrow{Y} = j(\kappa_1 + \kappa_2). \tag{3.166c}$$

Since $g''(x, x'; \lambda)$ is not an even function of either κ_1 or κ_2, branch points exist in the complex λ plane at $\lambda = -k_1^2$ and $\lambda = -k_2^2$. The argument of $\hat{\lambda} = \lambda + k_2^2$ is then restricted in accordance with $\mathrm{i}\kappa_1 < 0$, $\mathrm{i}\kappa_2 < 0$, as follows [see (3.150) and (3.160)]:

$$0 > \arg \hat{\lambda} > -2\pi, \quad 0 > \arg(\hat{\lambda} + h) > -2\pi, \quad h = k_1^2 - k_2^2, \tag{3.167}$$

with corresponding branch cuts along the real $\hat{\lambda}$ axis. Since g'' possesses no pole singularities on the branch of the Riemann surface for which $\mathrm{i}\kappa_1 < 0$ and $\mathrm{i}\kappa_2 < 0$, it is possible to find an appropriate contour of integration. Since the replacement of $\hat{\lambda}$ by $\hat{\lambda}e^{-j2\pi}$ in g'' yields g''^{*} [see (3.152c)], we may write

$$\begin{aligned}\delta(x - x') &= -\frac{1}{2\pi j}\oint_C g''(x, x'; \lambda)\, d\hat{\lambda} \\ &= -\frac{2}{\pi}\mathrm{i}\int_0^{\sqrt{h}} \xi_1 g''(x, x'; \xi_1^2 - k_1^2)\, d\xi_1 - \frac{2}{\pi}\mathrm{i}\int_0^{\infty} \xi g''(x, x'; \xi^2 - k_2^2)\, d\xi \end{aligned} \tag{3.168a}$$

$$= \int_0^{\sqrt{h}} \hat{\psi}_\Omega(\xi_1, x)\hat{\psi}^*_\Omega(\xi_1, x')\, d\xi_1 + \int_0^{\infty} \hat{\psi}'_\Omega(\xi, x)\hat{\psi}'^{*}_\Omega(\xi, x)\, d\xi, \quad -\infty < \begin{Bmatrix} x \\ x' \end{Bmatrix} < \infty \tag{3.168b}$$

where $\Omega = 1$ and $\Omega = 2$ correspond to $-\infty < (x \text{ or } x') < 0$ and $0 < (x$ or $x') < \infty$, respectively. For $0 < \xi < \infty$, one has the two mutually orthogonal sets

$$\hat{\psi}'_1(\xi, x) = \sqrt{\frac{1 - \overrightarrow{\Gamma}_1(\xi, 0)}{\pi}} \begin{Bmatrix} \cos \xi_1 x \\ \sqrt{\dfrac{\xi}{\xi_1}} \sin \xi_1 x \end{Bmatrix}, \quad -\infty < x < 0, \tag{3.169a}$$

$$\hat{\psi}'_2(\xi, x) = \sqrt{\frac{1 - \overrightarrow{\Gamma}_1(\xi, 0)}{\pi}} \begin{Bmatrix} \cos \xi x \\ \sqrt{\dfrac{\xi_1}{\xi}} \sin \xi x \end{Bmatrix}, \quad 0 < x < \infty, \tag{3.169b}$$

with

$$\xi_1 = \sqrt{\xi^2 + h} > 0, \quad \overrightarrow{\Gamma}_1(\xi, 0) = \frac{\xi_1 - \xi}{\xi_1 + \xi}. \tag{3.169c}$$

For $0 < \xi_1 < \sqrt{h}$ (i.e., $\xi = -j|\xi|$), the reflection coefficient $\overrightarrow{\Gamma}_1$ is complex and of unit magnitude; one has

$$\hat{\psi}_1(\xi_1, x) = \frac{1}{\sqrt{2\pi}}[e^{-j\xi_1 x} + \overrightarrow{\Gamma}_1(-j|\xi|, 0)e^{j\xi_1 x}], \quad -\infty < x < 0, \tag{3.170a}$$

$$\hat{\psi}_2(\xi_1, x) = \frac{1}{\sqrt{2\pi}}[1 + \overrightarrow{\Gamma}_1(-j|\xi|, 0)]e^{-|\xi|x}, \quad 0 < x < \infty. \tag{3.170b}$$

E modes (in x)
Characteristic Green's function:

$$g'(x, x'; \lambda) = \frac{\overleftarrow{I}_\Omega(x_<)\overrightarrow{I}_\Omega(x_>)}{j\omega\varepsilon_0 \overleftrightarrow{Z}(0)}, \tag{3.171a}$$

with

$$\overleftarrow{I}_1(x) = e^{j\kappa_1 x}, \quad \mathrm{i}\kappa_1 < 0, \quad \kappa_1^2 \equiv \xi_1^2 = k_1^2 + \lambda, \tag{3.171b}$$

$$\overrightarrow{I}_2(x) = e^{-j\kappa_2 x}, \quad \mathrm{i}\kappa_2 < 0, \quad \kappa_2^2 \equiv \xi^2 = k_2^2 + \lambda = \hat{\lambda}, \tag{3.171c}$$

$$\overrightarrow{I}_1(x) = \cos\kappa_1 x - j\frac{\bar{\varepsilon}_1\kappa_2}{\bar{\varepsilon}_2\kappa_1}\sin\kappa_1 x, \tag{3.171d}$$

$$\overleftarrow{I}_2(x) = \cos\kappa_2 x + j\frac{\bar{\varepsilon}_2\kappa_1}{\bar{\varepsilon}_1\kappa_2}\sin\kappa_2 x, \tag{3.171e}$$

$$j\omega\varepsilon_0 \overleftrightarrow{Z}(0) = j\left(\frac{\kappa_2}{\bar{\varepsilon}_2} + \frac{\kappa_1}{\bar{\varepsilon}_1}\right). \tag{3.171f}$$

Spectral representation of delta function for $-\infty < \begin{Bmatrix} x \\ x' \end{Bmatrix} < \infty$:

$$\begin{aligned}\bar{\varepsilon}(x)\delta(x - x') &= -\frac{1}{2\pi j}\oint_C d\hat{\lambda}\, g'(x, x'; \lambda) \\ &= \int_0^{\sqrt{h}} d\xi_1\, \hat{\Phi}_\Omega(\xi_1, x)\hat{\Phi}_\Omega^*(\xi_1, x') + \int_0^\infty d\xi\, \hat{\Phi}'_\Omega(\xi, x)\hat{\Phi}'_\Omega{}^*(\xi, x'),\end{aligned} \tag{3.172}$$

where C is the integration contour, and with $0 < \xi < \infty$,

$$\hat{\Phi}'_1(\xi, x) = \sqrt{\frac{\bar{\varepsilon}_2}{\pi}[1 + \overrightarrow{\Gamma}_1(\xi, 0)]}\begin{Bmatrix} \cos\xi_1 x \\ \sqrt{\dfrac{\xi\bar{\varepsilon}_1}{\xi_1\bar{\varepsilon}_2}}\sin\xi_1 x \end{Bmatrix}, \quad -\infty < x < 0, \tag{3.173a}$$

$$\hat{\Phi}'_2(\xi, x) = \sqrt{\frac{\bar{\varepsilon}_2}{\pi}[1 + \overrightarrow{\Gamma}_1(\xi, 0)]}\begin{Bmatrix} \cos\xi x \\ \sqrt{\dfrac{\xi_1\bar{\varepsilon}_2}{\xi\bar{\varepsilon}_1}}\sin\xi x \end{Bmatrix}, \quad 0 < x < \infty, \tag{3.173b}$$

with

$$\overrightarrow{\Gamma}_1(\xi, 0) = \frac{\xi - \xi_1(\bar{\varepsilon}_2/\bar{\varepsilon}_1)}{\xi + \xi_1(\bar{\varepsilon}_2/\bar{\varepsilon}_1)}, \quad \xi_1 = \sqrt{\xi^2 + h}. \tag{3.173c}$$

Also with $0 < \xi_1 < \sqrt{h}$

$$\hat{\Phi}_1(\xi_1, x) = \frac{1}{\sqrt{2\pi}} \left[e^{-j\xi_1 x} - \overrightarrow{\Gamma}_1(-j|\xi|, 0) e^{j\xi_1 x} \right], \quad -\infty < x < 0, \tag{3.174a}$$

$$\hat{\Phi}_2(\xi_1, x) = \frac{1}{\sqrt{2\pi}} \left[1 - \overrightarrow{\Gamma}_1(-j|\xi|, 0) \right] e^{-|\xi| x}, \quad 0 < x < \infty. \tag{3.174b}$$

References

[1] D. S. Jones, *Acoustic and Electromagnetic Waves.* Oxford, England: Clarendon Press, 1986.

[2] ——, *Methods in Electromagnetic Wave Propagation.* Oxford, England: Clarendon Press, 1987.

[3] J. A. Kong, *Electromagnetic Wave Theory.* Singapore: John Wiley & Sons, 1986.

[4] R. F. Harrington, "Origin and development of the method of moments for field computations," in *Computational Electromagnetic.* IEEE Press, 1992.

[5] R. Mittra (ed.), *Computer Techniques for Electromagnetics.* New York: Hemisphere Publishing Corporation, 1987.

[6] D. B. Davidson, *Computational electromagnetics for RF and Microwave Engineering.* Cambridge: Cambridge University Press, 2006.

[7] R. E. Collin, *Field Theory of Guided Waves.* New York: IEEE Press, 1991.

[8] L. P. Eisenhart, "Separable systems of stäckel," *Ann. Math.*, vol. 135, p. 284, 1934.

[9] J. A. Stratton, *Electromagnetic Theory.* New York, NY: McGraw-Hill, 1941.

[10] R. F. Harrington, *Time Harmonic Electromagnetic Fields.* New York: McGraw-Hill, 1961.

[11] L. B. Felsen, "Complexity architecture, phase space dynamics and problem–matched Green's functions," *Wave Motion*, vol. 34, pp. 243 –262, 2001.

[12] T. Rozzi and M. Mongiardo, *Open Electromagnetic Waveguides.* London: IEE, 1997.

[13] C. T. Tai, *Dyadics Green's Functions in Electromagnetic Theory.* Scranton, PA: Intext Educational Publishers, 1971.

[14] L. B. Felsen and N. Marcuvitz, *Radiation and Scattering of Waves.* Englewood Cliffs, NJ: Prentice Hall, 1973, Piscataway, NJ: IEEE Press (classic reissue), 1994.

4

Two–Dimensional Problems

I Introduction

In this chapter, the general concepts introduced in Chapter 2 SectionVI are illustrated on two examples:

- A parallel plate waveguide in a rectilinear y–independent domain, and
- Various axially invariant waveguides in a cylindrical coordinate domain.

The first example is very instructive because with a modest analytical effort allows to introduce several concepts for open and closed waveguides [1–3] and can be extended to dielectric waveguides [4] or stratified media. Also relevant techniques of applied mathematics find, in this case, immediate application [5–7].
The axially invariant waveguides in cylindrical coordinates illustrate the alternatives available when considering wave propagation in this coordinate system. Also in this case, with a limited analytical development, it is feasible to illustrate the relevant phenomenology [8–10]. These examples provide a simple yet effective introduction to the concepts that will be expressed in the next chapter, concerning the spherical wave expansion and its network interpretation.

II Electric Line Source in a PEC Parallel Plate Waveguide

We consider a parallel plate waveguide of height a, excited by a time-harmonic electric line source at location (x', z'), as shown in Figure 4.1. The scalar Green's function $G(x, z, x', z'; k_0)$ in the waveguide satisfies the time-harmonic wave equation

$$\left[\frac{\partial^2}{\partial x^2} + \frac{\partial^2}{\partial z^2} + k_0^2\right] G(x, z, x', z'; k_0) = -\delta(x - x')\delta(z - z'), \qquad (4.1)$$

where $k_0 = \omega / c_0$, ω is the radian frequency of the source and c_0 is the wave speed of the ambient medium in the waveguide. At the horizontal boundaries $x = 0$ and $x = a$, the PEC boundary conditions

$$G(0, z, x', z'; k_0) = G(a, z, x', z'; k_0) = 0 \qquad (4.2)$$

are imposed. A radiation condition, to be discussed in detail below, will be applied at $|z| \to \infty$. Since (4.1) is separable in x and z coordinates, the Green's function $G(x, z; x', z'; k_0)$ may be synthesized from the reduced one-dimensional problems in the x and z coordinates (see SectionV). Three approaches are possible:

- (1) G may be written as an eigenfunction expansion using the eigenfunctions of the reduced x-domain problem with coefficients that depend on z;
- (2) G may be written as an eigenfunction expansion using the eigenfunctions of the reduced z-domain problem with coefficients that depend on x; or
- (3) G may be written as an eigenfunction expansion using the two-dimensional eigenfunctions in the (x, z) domain with coefficients that are coordinate-independent.

The bounded x-domain will determine a discrete set of eigenfunctions in the variable x, while the unbounded z-domain will imply that the eigenfunctions in z form a continuous set. Before attacking the full two-dimensional problem, the reduced one-dimensional problems in the bounded x-domain, semi-infinite z-domain, and the bilaterally infinite z-domain will be considered first. The above solution strategies are explored, as is their usefulness with respect to physical insight, efficient numerical implementation and other considerations for the range of problem parameters of interest. Thus, this simplest of waveguide prototypes can serve as the setting for exploring many aspects that must be understood on an elementary level before they become "corrupted" by the more complicated phenomenology in more general waveguide environments.

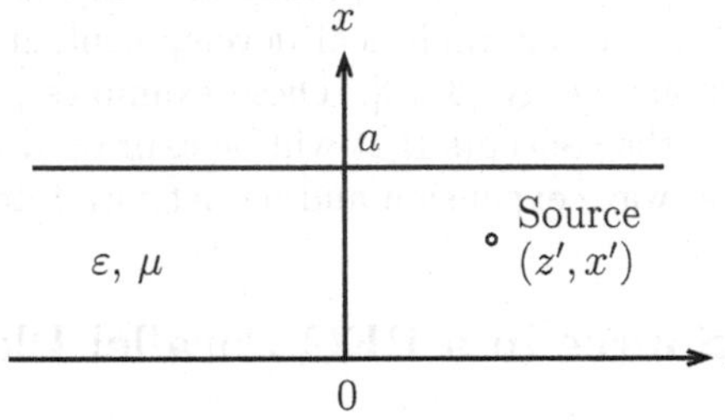

Fig. 4.1. Electric line source in a two-dimensional PEC parallel plate waveguide filled with a homogeneous dielectric.

II.1 Constituent One–Dimensional Problems: x-Domain

Eigenvalue Problem in the x-Domain

Equation (4.1) is coordinate separable, and the one-dimensional (reduced) homogeneous eigenvalue problem in the variable x is (see section V.1, and the corresponding reduction of (2.127) with $u = x$, $u_1 = 0$, $u_2 = a$, $p = w = 1$, $q = 0$, and with $\gamma_{1,2} = 0$ in (2.129))

$$\left(\frac{d^2}{dx^2} + \lambda_\alpha\right) f_\alpha(x) = 0, \quad 0 \le x \le a, \quad f_\alpha(0) = f_\alpha(a) = 0, \tag{4.3}$$

where λ_α is a separation constant which can be interpreted as the square of the x-component of the spatial wavenumber. The solution of (4.3) which satisfies the boundary condition $f_\alpha(0) = 0$ is

$$f_\alpha(x) = A_\alpha \sin \sqrt{\lambda_\alpha} x. \tag{4.4}$$

The boundary condition $f_\alpha(a) = 0$ requires

$$0 = A_\alpha \sin \sqrt{\lambda_\alpha} a, \tag{4.5}$$

which implies that

$$\lambda_\alpha = \left(\frac{\alpha\pi}{a}\right)^2, \quad \alpha = 1, 2, 3, \ldots. \tag{4.6}$$

The orthonormality condition

$$\int_0^a |f_\alpha|^2 \, dx = 1 \tag{4.7}$$

is now

$$A_\alpha^2 \int_0^a \sin^2 \sqrt{\lambda_\alpha} x \, dx = 1, \tag{4.8}$$

which gives the result

$$A_\alpha = \sqrt{\frac{2}{a}}. \tag{4.9}$$

The eigenfunctions f_α are therefore

$$f_\alpha(x) = \sqrt{\frac{2}{a}} \sin \frac{\alpha\pi x}{a}, \quad \alpha = 1, 2, 3, \ldots, \tag{4.10}$$

and the eigenvalues λ_α are given by (4.6) A sketch of the eigenfunction $f_\alpha(x)$ is shown in Figures 4.2. (4.6) and (4.10) constitute the classical solution of the eigenvalue problem in the x-domain.

Green's Function Problem for the x-Domain

The Green's function $g(x, x'; \lambda_x)$ for the reduced inhomogeneous problem in the variable x satisfies the equation

$$\left(\frac{d^2}{dx^2} + \lambda_x\right) g(x, x'; \lambda_x) = -\delta(x - x'), \qquad 0 \le x \le a, \tag{4.11}$$

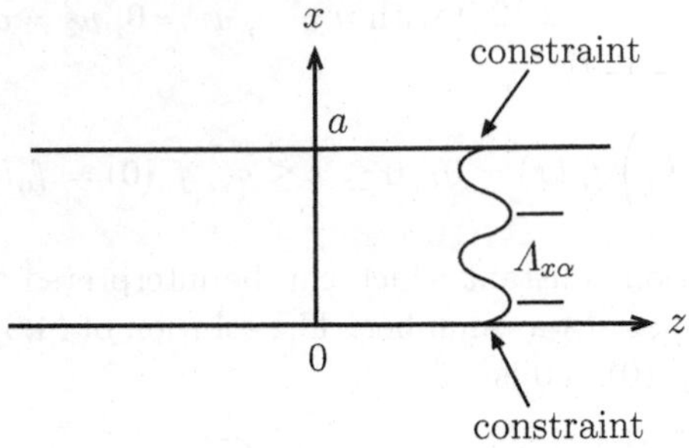

Fig. 4.2. Schematic representation of oscillatory (standing wave) eigenfunction $f_\alpha(x)$ in finite x-domain. The following correspondence holds: $\lambda_\alpha \to k_{x\alpha}^2$, $k_{x\alpha} = 2\pi/\Lambda_{x\alpha}$, where $\Lambda_{x\alpha}$ is the wavelength of the standing wave.

together with the boundary conditions

$$g(0, x'; \lambda_x) = g(a, x'; \lambda_x) = 0. \tag{4.12}$$

Here, λ_x is treated as a general complex parameter ((4.11) and (4.12) are a special case of (2.164) and (2.165)). The solution for $g(x, x'; \lambda_x)$ may be obtained by applying the general results derived in Section VI.2. As in Section II.1, $(u, u') = (x, x')$, $p = 1$, $q = 0$ and $w = 1$ in the general Sturm-Liouville operator given in (2.127). The functions $\overleftarrow{f}$ and $\overrightarrow{f}$ in (2.187) may be chosen as

$$\overleftarrow{f} = \sin \sqrt{\lambda_x} x \tag{4.13}$$

and

$$\overrightarrow{f} = \sin \sqrt{\lambda_x}(a - x). \tag{4.14}$$

The Wronskian of these two functions is

$$\begin{aligned} W &= p(x') \left[\overleftarrow{f} \frac{d\overrightarrow{f}}{dx} - \overrightarrow{f} \frac{d\overleftarrow{f}}{dx} \right]_{x=x'} \\ &= \sqrt{\lambda_x} \left[-\sin \sqrt{\lambda_x} x \cos \sqrt{\lambda_x}(a - x) - \sin \sqrt{\lambda_x}(a - x) \sin \sqrt{\lambda_x} x \right] \\ &= -\sqrt{\lambda_x} \sin \sqrt{\lambda_x} a, \end{aligned} \tag{4.15}$$

which, as predicted by the general result in (2.192), is independent of x'. According to (2.187), the Green's function $g(x, x'; \lambda_x)$ is therefore given by

$$g(x, x'; \lambda_x) = \frac{\sin \sqrt{\lambda_x} x_< \sin \sqrt{\lambda_x}(a - x_>)}{\sqrt{\lambda_x} \sin \sqrt{\lambda_x} a}. \tag{4.16}$$

A schematic representation of the function $g(x,x';\lambda_x)$ and of the functions $\overleftarrow{f}$ and $\overrightarrow{f}$ which comprise g is shown in Figure 4.3. It is evident from (4.16) that $g(x,x';\lambda_x)$ has simple poles at the real values $\lambda_x = \lambda_\alpha$ given by (4.6). The presence of $\sqrt{\lambda_x}$ in (4.16) suggests a branch point at $\lambda_x = 0$ in the complex λ-plane, but since $g(x,x';\lambda_x)$ is an even function of $\sqrt{\lambda_x}$, only even powers of $\sqrt{\lambda_x}$ (integral powers of λ_x) will appear in an expansion of g around $\lambda_x = 0$. This means that g is single-valued near $\lambda_x = 0$ (its phase will change by a multiple of 2π around a closed path containing the origin in the complex λ_x plane), and therefore there is no branch point at $\lambda_x = 0$.

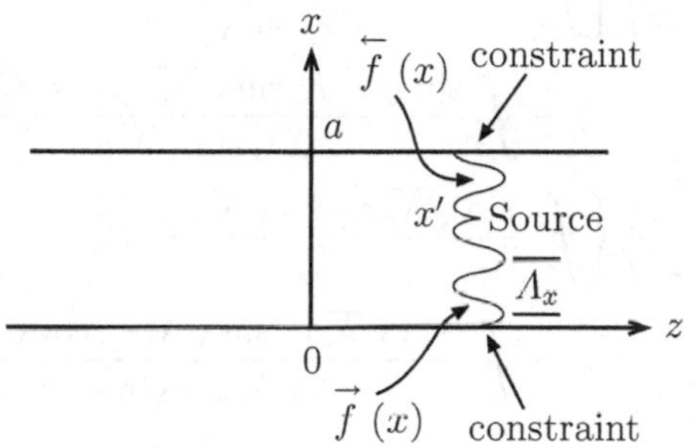

Fig. 4.3. Schematic representation of source-excited oscillatory wave functions in finite x-domain for arbitrary $\lambda_x \to k_x^2 = (2\pi/\Lambda_x)^2$.

The large λ_x behavior of g may be determined by writing the trigonometric functions in g in terms of complex exponentials, and retaining only the dominant exponential terms. For example, retaining only the dominant exponential order, we obtain in analogy to (2.197)

$$\sin\sqrt{\lambda_x}x = -\frac{1}{2j}\left(e^{-j\sqrt{\lambda_x}x} - e^{j\sqrt{\lambda_x}x}\right) \to O\left[e^{|\Im\sqrt{\lambda_x}|x}\right] \quad \text{as } |\lambda_x| \to \infty, \tag{4.17}$$

where the order notation $F(\lambda_x) \to O[G(\lambda_x)]$ as $|\lambda_x| \to \infty$ means that the ratio $F(\lambda_x)/G(\lambda_x)$ is bounded as $|\lambda_x| \to \infty$. Corresponding expressions can be written for $\sin\sqrt{\lambda_x}(a-x)$ and $\sin\sqrt{\lambda_x}a$. Substitution of these order estimates into (4.16) gives

$$\begin{aligned} g(x,x';\lambda_x) \to \frac{e^{|\mathrm{Im}\sqrt{\lambda_x}|x_<}e^{|\Im\sqrt{\lambda_x}|(a-x_>)}}{e^{|\Im\sqrt{\lambda_x}|a}} &= e^{|\Im\sqrt{\lambda_x}|(x_< - x_>)} \\ &= e^{|\Im\sqrt{\lambda_x}|(-|x-x'|)}. \end{aligned} \tag{4.18}$$

(4.18) shows explicitly that g decays exponentially as λ_x becomes large for all values of x except $x = x'$, thereby confirming the estimate established in (2.197), yielding no contribution over an integration path at $|\lambda| \to \infty$.

According to the general result given by (2.204), the integral of g around a closed contour C_x which encloses all poles $\lambda_x = \lambda_\alpha$ in the complex λ_x-plane gives

$$\delta(x - x') = \frac{1}{2\pi j} \oint_{C_x} g \, d\lambda_x. \tag{4.19}$$

For the present example, (4.19) becomes

$$\begin{aligned}
\delta(x - x') &= \frac{1}{2\pi j} \oint_{C_x} \frac{\sin\sqrt{\lambda_x}x_< \sin\sqrt{\lambda_x}(a - x_>)}{\sqrt{\lambda_x}\sin\sqrt{\lambda_x}a} \, d\lambda_x \\
&= \frac{1}{2\pi j} \left\{ \oint_{C_x} \frac{\sin\sqrt{\lambda_x}x_< \sin\sqrt{\lambda_x}a \cos\sqrt{\lambda_x}x_>}{\sqrt{\lambda_x}\sin\sqrt{\lambda_x}a} \, d\lambda_x \right. \\
&\qquad \left. - \oint_{C} \frac{\sin\sqrt{\lambda_x}x_< \sin\sqrt{\lambda_x}x_> \cos\sqrt{\lambda_x}a}{\sqrt{\lambda_x}\sin\sqrt{\lambda_x}a} \, d\lambda_x \right\} \\
&= \frac{1}{2\pi j} \left\{ \oint_{C_x} \frac{\sin\sqrt{\lambda_x}x_< \cos\sqrt{\lambda_x}x_>}{\sqrt{\lambda_x}} \, d\lambda_x \right. \\
&\qquad \left. - \oint_{C_x} \frac{\sin\sqrt{\lambda_x}x_< \sin\sqrt{\lambda_x}x_> \cos\sqrt{\lambda_x}a}{\sqrt{\lambda_x}\sin\sqrt{\lambda_x}a} \, d\lambda_x \right\}.
\end{aligned} \tag{4.20}$$

The integrand of the first term on the right-hand side of (4.20) has no singularities, and therefore (by Cauchy's Theorem) does not contribute. Again invoking Cauchy's Theorem, the second term may be evaluated by computing the residues at the simple poles λ_α given by (4.6). Expanding the denominator near $\lambda_x = \lambda_\alpha$ as $M(\lambda_x) = M(\lambda_\alpha) + (\lambda_x - \lambda_\alpha)(dM(\lambda_x)/d\lambda_x)_{\lambda_\alpha} + \ldots$, where $M(\lambda_x) = \sqrt{\lambda_x}\sin\sqrt{\lambda_x}a$, this calculation gives

$$\begin{aligned}
\delta(x - x') &= \frac{1}{2\pi j} \sum_\alpha 2\pi j \left\{ \frac{\sin\sqrt{\lambda_x}x_< \sin\sqrt{\lambda_x}x_> \cos\sqrt{\lambda_x}a}{\frac{d}{d\lambda_x}\left[\sqrt{\lambda_x}\sin\sqrt{\lambda_x}a\right]} \right\}_{\lambda_x = \lambda_\alpha} \\
&= \sum_\alpha \left\{ \frac{\sin\sqrt{\lambda_x}x_< \sin\sqrt{\lambda_x}x_> \cos\sqrt{\lambda_x}a}{\left[\frac{1}{2}\lambda_x^{-1/2}\sin\sqrt{\lambda_x}a + \sqrt{\lambda_x}\cos\sqrt{\lambda_x}a(a\frac{1}{2}\lambda_x^{-1/2})\right]} \right\}_{\lambda_x = \lambda_\alpha} \\
&= \sum_\alpha \frac{2}{a} \sin\sqrt{\lambda_\alpha}x_< \sin\sqrt{\lambda_\alpha}x_> \\
&= \sum_\alpha \frac{2}{a} \sin\frac{\alpha\pi x}{a} \sin\frac{\alpha\pi x'}{a} \\
&= \sum_\alpha f_\alpha(x) f_\alpha^*(x'),
\end{aligned} \tag{4.21}$$

which conforms with the general eigenfunction completeness expression in (2.154).

II.2 Problems in the z-Domain

Eigenvalue Problems in the Semi-Infinite z-Domain

As a first approach to an eigenvalue problem involving an unbounded domain, the semi-infinite problem

$$\left(\frac{d^2}{dz^2}+\lambda_\beta\right) f_\beta(z)=0, \quad 0\le z<\infty, \tag{4.22}$$

is considered, together with the boundary condition

$$f_\beta=0 \quad \text{at} \quad z=0. \tag{4.23}$$

The unbounded spatial domain makes this eigenvalue problem different from the finite domain eigenvalue problem considered in Chapter 2, Section II.2, because of the lack of a boundary condition at the imprecise "endpoint" $z\rightarrow\infty$.
While the solution of (4.22) which satisfies (4.23) is evidently

$$f_\beta(z)=C\sin\sqrt{\lambda_\beta}z, \tag{4.24}$$

the imprecise second boundary condition at $z\rightarrow\infty$ does not uniquely determine the allowable values of λ_β. In fact, *all* values of λ_β are "allowable," thereby implying that the λ_β are distributed continuously. However, this observation alone does not establish how the continuous set of eigenfunctions is normalized, nor what the discrete completeness statement in (2.154) becomes in the continuous limit.
The customary procedure for coping with this problem is to return to a finite domain $0<z<b$, and pass to the limit $b\rightarrow\infty$. Except for replacing a by b, α by β and x by z, the finite domain eigenvalue problem

$$\left(\frac{d^2}{dz^2}+\lambda_\beta\right) f_\beta(z)=0, \quad 0\le z\le b, \tag{4.25}$$

with the boundary conditions

$$f_\beta(0)=f_\beta(b)=0 \tag{4.26}$$

has been solved in Chapter 2, Section II.2, and the generalized eigenfunction completeness relation given by (2.154) becomes

$$\begin{aligned}\delta(z-z')&=\sum_\beta f_\beta(z)f_\beta^*(z')\\&=\sum_{\beta=0}^{\infty}\frac{2}{b}\sin(\sqrt{\lambda_\beta}z)\sin(\sqrt{\lambda_\beta}z')\\&=\sum_{\beta=0}^{\infty}\frac{2}{b}\sin\frac{\beta\pi z}{b}\sin\frac{\beta\pi z'}{b},\end{aligned} \tag{4.27}$$

where a term with index $\beta = 0$ has been included without changing the value of the sum. If the eigenvalues $\sqrt{\lambda_\beta} = \beta\pi/b$ are denoted by

$$\zeta_\beta = \frac{\beta\pi}{b}, \tag{4.28}$$

where the ζ_β is the z–domain spectral wavenumber, then the interval between eigenvalues is

$$\Delta\zeta_\beta = \zeta_{\beta+1} - \zeta_\beta = \frac{\pi}{b}. \tag{4.29}$$

With this notation, (4.27) can be written as

$$\delta(z - z') = \frac{2}{\pi}\sum_{\beta=0}^{\infty} \sin\left(\zeta_\beta z\right)\sin\left(\zeta_\beta z'\right)\Delta\,\zeta_\beta. \tag{4.30}$$

In the limit $b \to \infty$ (i.e., $\Delta\zeta_\beta \to 0$), the sum approaches an integral over the now continuous variable ζ, and (4.30) becomes

$$\delta(z - z') = \frac{2}{\pi}\int_0^\infty \sin\zeta z \sin\zeta z'\, d\zeta. \tag{4.31}$$

Equation (4.31) is the completeness relation synthesized by the function $\sin\zeta z$ of the continuous variable ζ over the interval $0 \le \zeta < \infty$, and may thus be taken as the defining equation for the orthonormal eigenfunctions of (4.22) in the semi-infinite z-domain. Comparing the first equality of (4.27) with (4.31) recalling that

$$\lambda_\beta \leftrightarrow \zeta^2 \tag{4.32}$$

leads to the identification

$$f_\beta(z) \leftrightarrow f_\zeta(z) = \sqrt{\frac{2}{\pi}}\sin\zeta z\,, \tag{4.33}$$

$$\sum_\beta \leftrightarrow \int_0^\infty d\zeta. \tag{4.34}$$

A schematic representation of the eigenfunction $f_\zeta(z)$ is shown in Figure 4.4. The eigenfunctions $\sin\zeta z$ are orthogonal, and the orthogonality relation can be derived in analogy with the discrete case by multiplying both sides of (4.31) by $\sin\zeta' z$ and integrating over z from 0 to ∞. Assuming the interchangeability of the orders of integration, this gives

$$\begin{aligned}\sin\zeta' z' &= \int_0^\infty \sin\zeta' z\left\{\frac{2}{\pi}\int_0^\infty \sin\zeta z\sin\zeta z'\, d\zeta\right\} dz\\ &= \int_0^\infty d\zeta\ \sin\zeta z'\left\{\frac{2}{\pi}\int_0^\infty dz\ \sin\zeta z\sin\zeta' z\right\},\end{aligned} \tag{4.35}$$

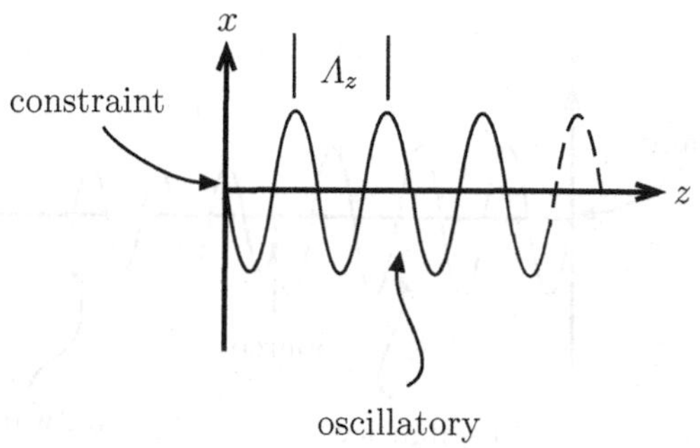

Fig. 4.4. Schematic representation of improper oscillatory eigenfunction $f_\zeta(z)$ in semi-infinite z-domain. $\zeta = 2\pi/\Lambda_z$, and *any* value of ζ is allowed.

from which it follows that

$$\frac{2}{\pi}\int_0^\infty dz\ \sin\zeta z \sin\zeta' z = \delta(\zeta - \zeta'). \tag{4.36}$$

Since the right-hand side of (4.36) is zero for $\zeta \neq \zeta'$, (4.36) is an orthogonality statement for the eigenfunctions $(\sqrt{2/\pi})\sin\zeta z$ and $(\sqrt{2/\pi})\sin\zeta' z$. The eigenfunctions *cannot* be *individually* normalized, however, since the integral on the left-hand side of (4.36) diverges for $\zeta = \zeta'$. For this reason, eigenfunctions in this category are called *improper.*

Equations (4.31) and (4.36) are the basis for the *Fourier sine transform* and its inverse, by which a function $F(z)$ which is defined over the domain $0 \leq z < \infty$ and which satisfies $F(0) = 0$ can be expressed as an integral over ζ of the basis functions $\sin\zeta z$. Starting from (4.31) and following the steps in (2.156)–(2.158), one finds the transform pair $F(z)$ and $\overline{F}(\zeta)$ which satisfy the equations

$$F(z) = \int_0^\infty \overline{F}(\zeta)\sin\zeta z\, d\zeta, \tag{4.37}$$

$$\overline{F}(\zeta) = \frac{2}{\pi}\int_0^\infty F(z)\sin\zeta z\, dz. \tag{4.38}$$

Because the introduction of the finite endpoint $z = b$ is artificial, the results obtained by taking the limit $b \to \infty$ should not depend on the boundary condition at $z = b$, which was here chosen as in (4.26). It can be shown that for infinite domain eigenvalue problems of the "limit point" type, to which the present problem belongs, the result of the limiting process is independent of the boundary condition at $z = b$. The difficulties encountered with the "direct" attack on the infinite domain eigenvalue (i.e., source-free) problem are avoided completely for the source-driven (Green's function) case discussed in the following section.

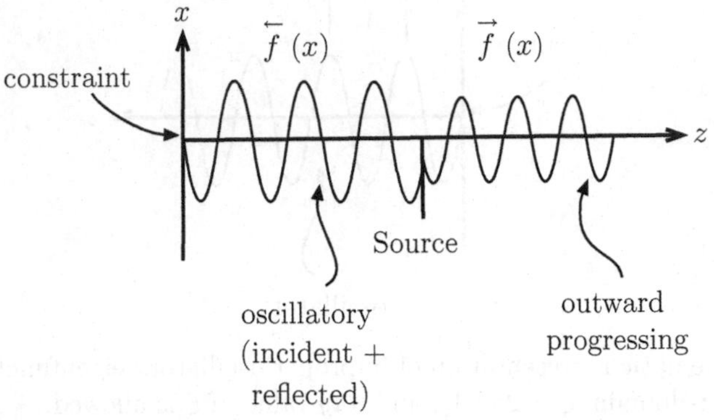

Fig. 4.5. Schematic representation of source-excited wave functions in semi-infinite z-domain. The squared wavenumber $\lambda_z = k_z^2 = (2\pi/\Lambda_z)^2$ is arbitrary.

Green's Function Problem in the Semi-Infinite z-Domain

As in Section II.1, the Green's function $g(z, z'; \lambda_z)$ for the reduced one-dimensional problem in the semi-infinite z domain satisfies the equation

$$\left(\frac{d^2}{dz^2} + \lambda_z\right) g(z, z', \lambda_z) = -\delta(z - z'), \ 0 \le z < \infty, \tag{4.39}$$

together with the boundary condition $g = 0$ at $z = 0$. For the behavior at $z \to \infty$, it suffices to require that g is bounded for *arbitrary* λ_z. The function $\overleftarrow{f}$ may again be chosen as

$$\overleftarrow{f} = \sin\sqrt{\lambda_z} z. \tag{4.40}$$

Boundedness at $z \to \infty$ implies that the function $\overrightarrow{f}$ behaves like

$$\overrightarrow{f} = e^{-j\sqrt{\lambda_z} z}, \quad \Im\sqrt{\lambda_z} < 0 \tag{4.41}$$

i.e. $0 \ge \arg\sqrt{\lambda_z} > -\pi$, on the "upper sheet" of the two-sheeted Riemann surface in the complex λ_z-plane. This condition ensures that $\overrightarrow{f}$ *decays* like $\exp\left[-|\Im\sqrt{\lambda_z} z|\right]$ with increasing $|\lambda_z|$ or z. The mathematical boundedness condition in (4.41) is consistent with the physically motivated "radiation condition" which requires that waves propagate outward from the source region toward $z \to \infty$, i.e. $exp^{-j\sqrt{\lambda_z} z}$ for the assumed $e^{j\omega t}$ time dependence. Complex λ_z implies dissipation in the propagation medium and a decaying wavefield as in (4.41). The Wronskian of the functions $\overleftarrow{f}$ and $\overrightarrow{f}$ defined by (4.40) and (4.41) is

$$
\begin{aligned}
W &= \left\{ \overleftarrow{f}\frac{d\overrightarrow{f}}{dz} - \overrightarrow{f}\frac{d\overleftarrow{f}}{dz} \right\}_{z=z'} \\
&= -\left\{ \left[\frac{e^{-j\sqrt{\lambda_z}z} + e^{j\sqrt{\lambda_z}z}}{2j} \right] (\sqrt{\lambda_z}) e^{-j\sqrt{\lambda_z}z} \right. \\
&\quad \left. - e^{-j\sqrt{\lambda_z}z}\sqrt{\lambda_z}\left[\frac{e^{-j\sqrt{\lambda_z}z} + e^{j\sqrt{\lambda_z}z}}{2} \right] \right\}_{z=z'} \\
&= -\sqrt{\lambda_z}.
\end{aligned}
\tag{4.42}
$$

According to (2.187), the Green's function is therefore

$$
g(z, z'; \lambda) = \frac{\sin\sqrt{\lambda_z}z_{<} e^{-j\sqrt{\lambda_z}z_{>}}}{\sqrt{\lambda_z}}. \tag{4.43}
$$

Figure 4.6 shows a schematic representation, for real λ_z, of the real part of the Green's function $g(z, z'; \lambda_z)$ in the semi-infinite z-domain, together with the functions $\overrightarrow{f}$ and $\overleftarrow{f}$.

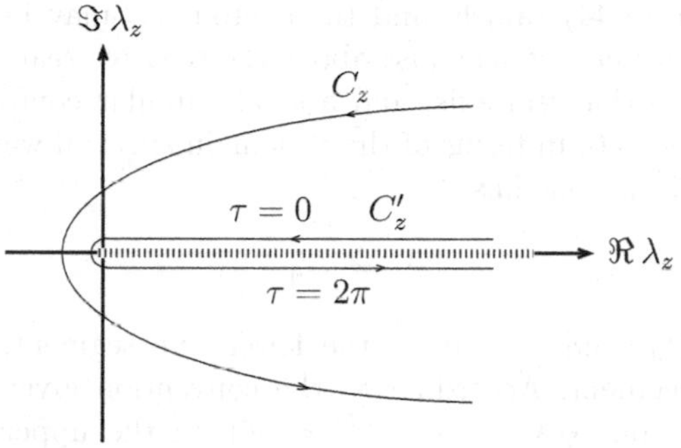

Fig. 4.6. Complex λ_z-plane with spectral branch cut.

To relate the Green's function given by (4.43) to the eigenfunctions of the associated sourceless problem, g is integrated over a contour C which encloses all the singularities of g in the complex λ_z-plane. The Green's function given by (4.43) has no poles, but does have a branch point at $\lambda_z = 0$ (i.e., g is multivalued on any closed contour that encircles $\lambda_z = 0$; note that this Green's function is *not* an even function of $\sqrt{\lambda_z}$). In order to define the function $\sqrt{\lambda_z}$ uniquely, a two-sheeted Riemann surface with a branch cut is introduced in the complex λ_z-plane. The branch cut must extend from the branch point to infinity. Passing through the cut once grants access from the upper to the lower sheet in the extended complex λ-plane, and passing through the cut twice re-enters the upper sheet.

As stated previously, the complex square root is defined here with the condition $0 \geq \arg\sqrt{\lambda_z} > -\pi$ on the upper sheet, thereby rendering g convergent on the entire upper sheet. Accordingly, the branch cut is chosen so that $\Im\sqrt{\lambda_z} = 0$ along the cut; this places the branch cut on the positive $\Re\lambda_z$ axis, as shown in Figure 4.6 (note that the changeover from the branches (or sheets) where $\Im\sqrt{\lambda_z} < 0$ and $\Im\sqrt{\lambda_z} > 0$, respectively, occurs along the contour $\Im\sqrt{\lambda_z} = 0$, which thereby defines the "spectral branch cut"). The generalized completeness theorem applied to the Green's function in (4.43) is now

$$\begin{aligned} \delta(z-z') &= \frac{1}{2\pi j}\oint_{C_z} g(z,z';\lambda_z)\,d\lambda_z \\ &= \frac{1}{2\pi j}\oint_{C_z} \frac{\sin\sqrt{\lambda_z}z_< e^{-j\sqrt{\lambda_z}z_>}}{\sqrt{\lambda_z}}\,d\lambda_z, \end{aligned} \tag{4.44}$$

where the contour C_z is shown in Figure 4.6. As $|\lambda_z| \to \infty$ in the complex plane, g behaves like

$$g \to e^{|\Im\sqrt{\lambda_z}|z_<} e^{-|\Im\sqrt{\lambda_z}|z_>} = e^{-|\Im\sqrt{\lambda_z}|(z_>-z_<)} = e^{-|\Im\sqrt{\lambda_z}||z-z'|}, \tag{4.45}$$

whence g decays as $|\lambda_z| \to \infty$ for $z \neq z'$. Therefore the contributions at infinity to the integral in (4.44) vanish, and the contour C_z may be deformed into the contour C_z' with a line segment just above the positive real axis, a line segment just below the positive real axis, and a small circular contour surrounding the branch point at $\lambda_z = 0$. In terms of the z–domain spectral wavenumber, $\zeta = \sqrt{\lambda_z}$ written in polar form, one has

$$\lambda_z = |\zeta^2| e^{-j\tau}, \tag{4.46}$$

with $\tau = \arg(\zeta^2)$, and $\tau = 0$ on the lower line segment, and $\tau = -2\pi$ on the upper line segment. According to the convention given previously for the square root function, $\sqrt{\lambda_z} = |\zeta| e^{-j(0)} = |\zeta|$ on the upper line segment, and $\sqrt{\lambda_z} = |\zeta| e^{-j\pi} = -|\zeta|$ on the lower line segment. It can be shown that the contribution from the small circular contour around $\lambda_z = 0$ vanishes in the limit as the radius of the circle approaches zero, so that (4.44) becomes

$$\begin{aligned} \delta(z-z') &= \frac{1}{2\pi j}\int_\infty^0 \frac{\sin\zeta z_< e^{-j\zeta z_>}}{\zeta}\,d\zeta^2 + \frac{1}{2\pi j}\int_0^\infty \frac{\sin(-\zeta z_<) e^{j\zeta z_>}}{-\zeta}\,d\zeta^2 \\ &= -\frac{1}{2\pi j}\int_0^\infty \sin\zeta z_< e^{-j\zeta z_>} 2d\zeta + \frac{1}{2\pi j}\int_0^\infty \sin\zeta z_< e^{j\zeta z_>}\,2d\zeta \\ &= -\frac{1}{2\pi j}\int_0^\infty \sin\zeta z_< \left(e^{-j\zeta z_>} - e^{j\zeta z_>}\right) 2d\zeta\,. \end{aligned} \tag{4.47}$$

From this we obtain

$$\delta(z-z') = \frac{2}{\pi}\int_0^\infty \sin\zeta z \sin\zeta z' \, d\zeta, \tag{4.48}$$

which is precisely the eigenfunction completeness relation. Thus, the eigenfunction completeness relation and the identification of the homogeneous eigenfunctions themselves can be derived from integration of the Green's function for the source-driven problem, thereby providing a *direct* alternative to the process of first solving the homogeneous problem in a finite domain and then considering the limit as the finite domain is allowed to become infinite.

Eigenvalue Problem in the Bilaterally Infinite z-Domain

The eigenvalue problem in the bilaterally infinite z-domain may again be approached by first considering the finite domain problem

$$\left(\frac{d^2}{dz^2} + \lambda_\beta\right) f_\beta(z) = 0, \quad -b/2 < z < b/2, \quad f_\beta(-b/2) = f_\beta(b/2) = 0. \tag{4.49}$$

Results for the eigenvalue problem defined over the domain $-\infty < z < \infty$ will be obtained by letting $b \to \infty$.
The solution of (4.49) may be obtained by using the results of Section with a simple shift in the origin of z; the eigenvalues are

$$\zeta_\beta = \left(\frac{\beta\pi}{b}\right)^2, \tag{4.50}$$

and the normalized eigenfunctions are

$$f_\beta(z) = \sqrt{\frac{2}{b}} \sin\frac{\beta\pi}{b}(z + \frac{b}{2}). \tag{4.51}$$

The eigenfunction completeness relation is now

$$\begin{aligned}\delta(z-z') &= \sum_\beta f_\beta(z) f_\beta^*(z') \\ &= \sum_{\beta=1}^\infty \frac{2}{b} \sin\frac{\beta\pi}{b}(z+\frac{b}{2}) \sin\frac{\beta\pi}{b}(z'+\frac{b}{2}).\end{aligned} \tag{4.52}$$

In contrast to the choice of coordinates in (4.3), the placement of the origin in the center of the finite interval, as in (4.49), highlights eigensolutions with even and odd symmetry. The right-hand side of (4.52) is even in β, and a term with $\beta = 0$ may again be added without changing the value of the sum, so that (4.52) can be written as

$$\delta(z-z') = \sum_{\beta=-\infty}^\infty \frac{1}{b} \sin\frac{\beta\pi}{b}(z+\frac{b}{2}) \sin\frac{\beta\pi}{b}(z'+\frac{b}{2}). \tag{4.53}$$

Expanding the trigonometric sums in (4.53) gives

$$\delta(z-z') = \sum_{\beta=-\infty}^{\infty} \frac{1}{b}\left\{\sin\frac{\beta\pi z}{b}\sin\frac{\beta\pi z'}{b}\cos^2\frac{\beta\pi}{2} + \cos\frac{\beta\pi z}{b}\cos\frac{\beta\pi z'}{b}\sin^2\frac{\beta\pi}{2}\right\}, \tag{4.54}$$

where use has been made of the fact that $\sin(\beta\pi/2)\cos(\beta\pi/2)$ is zero for all integers β. Also,

$$\cos^2\frac{\beta\pi}{2} = \begin{cases} 1, \beta \text{ even} \\ 0, \beta \text{ odd} \end{cases} \tag{4.55}$$

and

$$\sin^2\frac{\beta\pi}{2} = \begin{cases} 1, \beta \text{ odd} \\ 0, \beta \text{ even}, \end{cases} \tag{4.56}$$

so that (4.54) can be written as

$$\delta(z-z') = \sum_{\beta \text{ even}} \frac{1}{b}\sin\frac{\beta\pi z}{b}\sin\frac{\beta\pi z'}{b} + \sum_{\beta \text{ odd}} \frac{1}{b}\cos\frac{\beta\pi z}{b}\cos\frac{\beta\pi z'}{b}. \tag{4.57}$$

By making the change of variables

$$\beta = 2\eta\,, \qquad \beta \text{ even}\,, \tag{4.58}$$

$$\beta = 2\gamma + 1\,, \qquad \beta \text{ odd}\,, \tag{4.59}$$

(4.57) becomes

$$\delta(z-z') = \sum_{\eta=-\infty}^{\infty} \frac{1}{b}\sin\frac{2\eta\pi z}{b}\sin\frac{2\eta\pi z'}{b} + \sum_{\gamma=-\infty}^{\infty} \frac{1}{b}\cos\frac{(2\gamma+1)\pi z}{b}\cos\frac{(2\gamma+1)\pi z'}{b}, \tag{4.60}$$

which with the notation

$$\zeta_\eta = \frac{2\eta\pi}{b}\,, \tag{4.61}$$

$$\zeta_\gamma = \frac{(2\gamma+1)\pi}{b}\,, \tag{4.62}$$

$$\Delta\zeta_\eta = \Delta\zeta_\gamma = \frac{2\pi}{b} \tag{4.63}$$

becomes in turn

$$\delta(z-z') = \frac{1}{2\pi}\sum_{\eta=-\infty}^{\infty} \sin\zeta_\eta z \sin\zeta_\eta z' \Delta\,\zeta_\eta + \frac{1}{2\pi}\sum_{\gamma=-\infty}^{\infty} \cos\zeta_\gamma z \cos\zeta_\gamma z' \Delta\,\zeta_\gamma. \tag{4.64}$$

As $b \to \infty$, both sums in (4.64) become integrals with respect to the continuous variable ζ over the range $-\infty < \zeta < \infty$ and (4.64) becomes

$$\begin{aligned}\delta(z-z') &= \frac{1}{2\pi}\int_{-\infty}^{\infty}(\sin\zeta z\sin\zeta z' + \cos\zeta z\cos\zeta z')\,d\zeta \\ &= \frac{1}{\pi}\int_0^{\infty}\sin\zeta z\sin\zeta z'\,d\zeta + \frac{1}{\pi}\int_0^{\infty}\cos\zeta z\cos\zeta z'\,d\zeta \\ &= \frac{1}{2\pi}\int_{-\infty}^{\infty}\cos\zeta(z-z')\,d\zeta. \end{aligned} \tag{4.65}$$

Since $\sin\zeta(z-z')$ is odd in ζ and will not contribute to an integral over a symmetric range of ζ, (4.65) can also be written as

$$\begin{aligned}\delta(z-z') &= \frac{1}{2\pi}\int_{-\infty}^{\infty}[\cos\zeta(z-z') - j\sin\zeta(z-z')]\,d\zeta \\ &= \frac{1}{2\pi}\int_{-\infty}^{\infty}e^{-j\zeta(z-z')}\,d\zeta. \end{aligned} \tag{4.66}$$

The first equality shows that the completeness relations in (4.65) and (4.66) contain alternative eigenfunction sets for the bilaterally infinite z-domain. In (4.66), the normalized improper eigenfunctions are $\sqrt{1/2\pi}e^{-j\zeta z}$ over the interval $-\infty < \zeta < \infty$. In particular, comparing (4.66) and (4.52), one has the identifications

$$\lambda_\beta \leftrightarrow \zeta^2\,, \tag{4.67}$$

$$f_\beta(z) \leftrightarrow f_\zeta(z) = \sqrt{1/2\pi}e^{-j\zeta z}\,, \tag{4.68}$$

$$f_\beta^*(z') \leftrightarrow f_\zeta^*(z') = \sqrt{1/2\pi}e^{j\zeta z'}\,, \tag{4.69}$$

$$\sum_\beta \leftrightarrow \int_{-\infty}^{\infty} d\zeta\,. \tag{4.70}$$

Returning to the second line in (4.65), on the other hand, one has the symmetric-antisymmetric eigenfunction decomposition over the semi-infinite domain $0 < \zeta < \infty$, which comprises *standing* (oscillatory) waves; (4.66) emphasizes *traveling* (progressive) waves.

As before, the orthogonality relation for the improper eigenfunctions $\sqrt{1/2\pi}e^{-j\zeta z}$ is derived by multiplying both sides of (4.66) by $e^{j\zeta' z}$ and integrating over z from $-\infty$ to ∞. This gives

$$\begin{aligned} e^{j\zeta' z'} &= \int_{-\infty}^{\infty} e^{j\zeta' z}\left\{\frac{1}{2\pi}\int_{-\infty}^{\infty} e^{-j\zeta(z-z')}\,d\zeta\right\}dz \\ &= \int_{-\infty}^{\infty} e^{j\zeta z'}\left\{\frac{1}{2\pi}\int_{-\infty}^{\infty} e^{-j(\zeta-\zeta')z}\,dz\right\}d\zeta, \end{aligned} \tag{4.71}$$

from which it follows that

$$\frac{1}{2\pi}\int_{-\infty}^{\infty} e^{-j(\zeta-\zeta')z}\,dz = \delta(\zeta-\zeta'). \tag{4.72}$$

Similar relations can be derived for the symmetric and antisymmetric eigenfunction sets in the second line of (4.65).

Equations (4.66) and (4.72) are the basis for the *Fourier transform*, by which a function $F(z)$ defined over the domain $-\infty < z < \infty$ can be written as a continuous sum (integral) over ζ of the basis functions $\sqrt{1/2\pi e^{-j\zeta z}}$. Starting from (4.66) and again following the steps in (2.156)–(2.158), the Fourier transform and its inverse are obtained in the form

$$F(z) = \frac{1}{2\pi}\int_{-\infty}^{\infty} \overline{F}(\zeta)e^{-j\zeta z}\,d\zeta, \tag{4.73}$$

$$\overline{F}(\zeta) = \int_{-\infty}^{\infty} F(z)e^{j\zeta z}\,dz. \tag{4.74}$$

Green's Function Problem in the Bilaterally Infinite z-Domain

The Green's function $g(z, z'; \lambda_z)$ for the one-dimensional problem in the bilaterally infinite z domain satisfies the equation

$$\left(\frac{d^2}{dx^2} + \lambda_z\right) g(z, z'; \lambda_z) = -\delta(z - z'), \quad -\infty < z < \infty. \tag{4.75}$$

As in Section II.2, the functions $\overleftarrow{f}$ and $\overrightarrow{f}$ used to construct the Green's function are chosen so as to yield bounded solutions (satisfy radiation conditions) as $z \to \pm\infty$, i.e.,

$$\overleftarrow{f} = e^{j\sqrt{\lambda_z}z}, \tag{4.76}$$

$$\overrightarrow{f} = e^{-j\sqrt{\lambda_z}z} \tag{4.77}$$

with the Wronskian

$$\begin{aligned} W &= \left\{ \overleftarrow{f}\frac{d\overrightarrow{f}}{dz} - \overrightarrow{f}\frac{d\overleftarrow{f}}{dz} \right\}_{z=z'} \\ &= -\left\{ e^{j\sqrt{\lambda_z}z}(j\sqrt{\lambda_z})e^{-j\sqrt{\lambda_z}z} - e^{-j\sqrt{\lambda_z}z}(j\sqrt{\lambda_z})e^{j\sqrt{\lambda_z}z} \right\}_{z=z'} \\ &= -2j\sqrt{\lambda_z}. \end{aligned} \tag{4.78}$$

The schematic representation of the wave process is as in Figure 4.5, except that $\overleftarrow{f}$ is outward progressing towards $z \to -\infty$ in accord with the radiation condition. Using the general result in (2.187), the Green's function is

$$g(z, z'; \lambda_z) = \frac{e^{-j\sqrt{\lambda_z}z_>}e^{j\sqrt{\lambda_z}z_<}}{2j\sqrt{\lambda_z}} = \frac{e^{-j\sqrt{\lambda_z}|z-z'|}}{2j\sqrt{\lambda_z}}, \tag{4.79}$$

yielding the generalized completeness relation

$$\delta(z - z') = \frac{1}{2\pi j} \oint_{C_z} g(z, z'; \lambda_z) \, d\lambda_z = \frac{1}{2\pi j} \oint_{C_z} \frac{e^{-j\sqrt{\lambda_z}|z-z'|}}{2j\sqrt{\lambda_z}} \, d\lambda_z, \tag{4.80}$$

where C_z is a contour which encloses in the positive sense all the singularities of g in the complex λ_z-plane. The Green's function given by (4.79) has a branch point at $\lambda_z = 0$ and no other singularities. As in Figure 4.6, the (spectral) branch cut is chosen on the positive $\Re\,\lambda_z$ axis. The contour C_z may be deformed into the contour C'_z as in Section II.2. Proceeding as in (4.47)-(4.48), (4.80) becomes

$$\delta(z - z') = \frac{1}{2\pi j} \int_{\infty}^{0} \frac{e^{-j\zeta|z-z'|}}{2j\zeta} \, d\zeta^2 - \frac{1}{2\pi j} \int_{0}^{\infty} \frac{e^{j\zeta|z-z'|}}{2j\zeta} \, d\zeta^2, \tag{4.81}$$

which reduces to

$$\delta(z - z') = \frac{1}{2\pi} \int_0^\infty \left(e^{-j\zeta|z-z'|} + e^{j\zeta|z-z'|} \right) d\zeta = \frac{1}{\pi} \int_0^\infty \cos\zeta|z - z'| \, d\zeta. \tag{4.82}$$

Since the cosine is an even function, the absolute value in the integrand can be removed, giving

$$\begin{aligned} \delta(z - z') &= \frac{1}{\pi} \int_0^\infty \cos\zeta(z - z') \, d\zeta = \frac{1}{2\pi} \int_{-\infty}^{\infty} \cos\zeta(z - z') \, d\zeta \\ &= \frac{1}{2\pi} \int_{-\infty}^{\infty} e^{-j\zeta(z-z')} \, d\zeta. \end{aligned} \tag{4.83}$$

(4.83) is the eigenfunction completeness relation in Section II.2.

II.3 Two-Dimensional Waveguide:(Finite x)–(Bilaterally Infinite z)–Domain

Eigenvalue Problem

The two-dimensional eigenvalue problem for the two-dimensional geometry in Figure 4.1 defines eigenfunctions which span the entire (x, z) domain, in contrast to the reduced one-dimensional eigenvalue problems in Section II.2 and II.2 for the x and z domains, respectively.
The two-dimensional eigenvalue problem is defined by the equation

$$\left(\frac{\partial^2}{\partial x^2} + \frac{\partial^2}{\partial z^2} + \lambda_\nu \right) F_\nu(x, z) = 0, \quad 0 \le x \le a, \ -\infty < z < \infty \tag{4.84}$$

with boundary conditions $F_\nu(x, z) = 0$ at $x = 0, a$ and indefinite boundary conditions at $|z| \to \infty$ (see Sections II.2 and II.2). Since the problem in eqn. (4.84) is coordinate–separable, the solution can be synthesized in terms of the one-dimensional orthonormal eigensets $\{f_\alpha(x)\}$, $\{f_\zeta(z)\}$ in (4.10) and (4.68), respectively. Thus, the eigenfunctions F_ν become

$$F_\nu \to F_{\alpha,\zeta}(x,z) = f_\alpha(x) f_\zeta(z) = \sqrt{\frac{2}{a}}\frac{1}{\sqrt{2\pi}} \sin\frac{\alpha\pi x}{a} e^{-j\zeta z}, \quad \nu \to (\alpha, \zeta) \qquad (4.85)$$

where $\alpha = 1, 2, 3, \ldots$, $-\infty < \zeta < \infty$, and the eigenvalues λ_ν in (4.84) become

$$\lambda_\nu \to \lambda_{\alpha,\zeta} = (\alpha\pi/a)^2 + \zeta^2. \qquad (4.86)$$

Accordingly, the two-dimensional completeness and normalization relations are given by

$$\delta(x-x')\delta(z-z') = \sum_\alpha \int_{-\infty}^{\infty} d\zeta\, F_{\alpha,\zeta}(x,z) F^*_{\alpha,\zeta}(x',z'), \qquad (4.87)$$

$$\int_0^a dx \int_{-\infty}^{\infty} dz\, F_{\alpha,\zeta}(x,z) F^*_{\overline{\alpha},\overline{\zeta}}(x,z) = \delta(\zeta - \overline{\zeta})\delta_{\alpha,\overline{\alpha}}\,. \qquad (4.88)$$

Green's Function Problem

The results in the previous sections permit synthesis of the two-dimensional Green's function for the original waveguide in Figure 4.1. Repeating (4.1) and (4.2), the two-dimensional Green's function $G(x, z; x', z'; k_0)$ satisfies the equation

$$\left(\frac{\partial^2}{\partial x^2} + \frac{\partial^2}{\partial z^2} + k_0^2\right) G(x,z;x',z';k_0) = -\delta(x-x')\delta(z-z') \qquad (4.89)$$

with boundary conditions

$$G = 0 \quad \text{at} \quad x = 0, a \qquad (4.90)$$

and the radiation condition at $|z| \to \infty$.

Eigenfunction Expansion in the x-Domain

To construct a representation for $G(x, z; x', z'; k_0)$ using the eigenfunctions of the reduced x-domain problem, $G(x, z; x', z'; k_0)$ is written as

$$G = \sum_\alpha A_\alpha(z, z', x') f_\alpha(x), \qquad (4.91)$$

where

$$f_\alpha(x) = \sqrt{\frac{2}{a}} \sin\frac{\alpha\pi x}{a} \qquad (4.92)$$

are the eigenfunctions of the finite x-domain problem discussed in Section II.2, and the coefficients A_α are to be determined. In view of the (x, z) separability and the (x, x') symmetry exhibited in (4.89), it is suggestive to reduce the coefficients A_α as

$$A_\alpha(z, z', x') = g_z(z, z'; \alpha) f_\alpha^*(x') \tag{4.93}$$

so that G becomes

$$G = \sum_\alpha g_z(z, z'; \alpha) f_\alpha^*(x') f_\alpha(x). \tag{4.94}$$

Using (4.3) and (4.6), the second derivative of G with respect to x is

$$\begin{aligned} \frac{\partial^2}{\partial x^2} G &= \sum_\alpha g_z(z, z'; \alpha) f_\alpha^*(x') \frac{d^2}{dx^2} f_\alpha(x) \\ &= \sum_\alpha g_z(z, z'; \alpha) f_\alpha^*(x')(-\lambda_\alpha f_\alpha(x)) \end{aligned} \tag{4.95}$$

with the x-domain eigenvalues λ_α given by

$$\lambda_\alpha = \left(\frac{\alpha\pi}{a}\right)^2. \tag{4.96}$$

The second derivative of G with respect to z is

$$\frac{\partial^2}{\partial z^2} G = \sum_\alpha \frac{d^2}{dz^2} g_z(z, z'; \alpha) f_\alpha^*(x') f_\alpha(x). \tag{4.97}$$

(4.89) now becomes

$$\sum_\alpha \left\{ \frac{d^2}{dz^2} g_z + k_0^2 - \lambda_\alpha \right\} f_\alpha(x) f_\alpha^*(x') = -\delta(x - x')\delta(z - z'), \tag{4.98}$$

which upon using (4.21) becomes

$$\sum_\alpha \left\{ \frac{d^2}{dz^2} g_z + k_0^2 - \lambda_\alpha \right\} f_\alpha(x) f_\alpha^*(x') = -\sum_\alpha f_\alpha(x) f_\alpha^*(x') \delta(z - z'). \tag{4.99}$$

Equating the coefficients of the orthogonal functions $f_\alpha(x)$, (4.99) implies that g_z satisfies the equation

$$\frac{d^2}{dz^2} g_z(z, z'; \alpha) + (k_0^2 - \lambda_\alpha) g_z(z, z'; \alpha) = -\delta(z - z'), \tag{4.100}$$

subject to the radiation condition at $|z| \to \infty$. (4.100) is the one-dimensional Green's function equation discussed in Section II.2, and the solution of (4.100) which satisfies the radiation condition at $|z| \to \infty$ is (see (4.79))

$$g_z(z, z'; \alpha) = \frac{e^{-j\sqrt{k_0^2 - \lambda_\alpha}|z - z'|}}{2j\sqrt{k_0^2 - \lambda_\alpha}}. \tag{4.101}$$

As before, the square root function in (4.101) is defined so that $0 \le \arg\sqrt{\ } < \pi$. The series expansion for $G(x, z; x', z'; k_0)$ in (4.94) is now

$$G = \sum_{\alpha} \frac{e^{-j\sqrt{k_0^2-\lambda_\alpha}|z-z'|}}{2j\sqrt{k_0^2-\lambda_\alpha}} \frac{2}{a} \sin\sqrt{\lambda_\alpha}x \sin\sqrt{\lambda_\alpha}x', \tag{4.102}$$

with the eigenvalues λ_α given by (4.96). This expression for G is written in terms of oscillatory eigenfunctions (modes) in the x-cross section, and emphasizes traveling waves along z. For schematic representation of these combined wave processes, see Figure 4.4 and Figure 4.7, modified as indicated after (4.78).

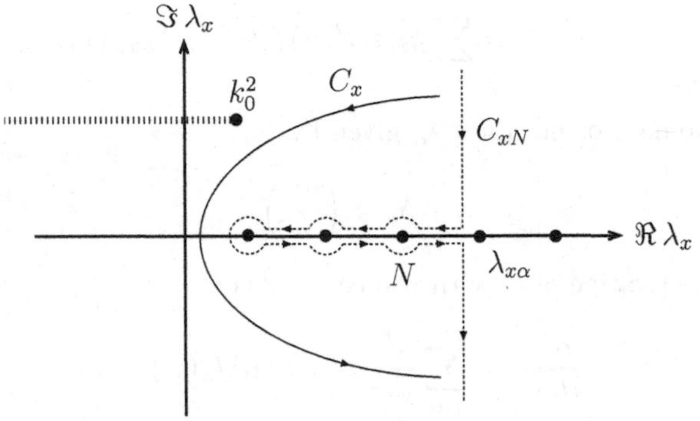

Fig. 4.7. Contours in the λ_x-plane.

The guided mode series in (4.102) is useful and phenomenologically meaningful when the number of "important" modes is not too large. The number of important modes at sufficiently long ranges $|z - z'|$ from the source is controlled by the exponential term (the z-modal propagator) in (4.101), which decays when $\lambda_\alpha > k_0^2 = \omega^2/c^2$ since, then, $\sqrt{k_0^2 - \lambda_\alpha} = -j|\sqrt{\lambda_\alpha - k_0^2}|$. Thus, the downrange modes are controlled by their cutoff frequencies $\omega_{c\alpha} = \alpha\pi c_0/a$, and they are "filtered out" by the waveguide whenever $\omega < \omega_{c\alpha}$, i.e. at "low enough" operating frequencies. However, this filtering takes place only if $|z - z'|$ is sufficiently large. As $z \to z'$, the exponential propagator for modes with $\omega_{c\alpha} > \omega$ is weakly damped, and in the cross section $z = z'$ of the source, the damping *disappears* altogether. Thus, alternative formulations may offer a more attractive option.

Eigenfunction Expansion in the z-Domain

An alternative representation for the two-dimensional Green's function $G(x, z; x', z'; k_0)$ is obtained by expanding G in eigenfunctions of the reduced z-domain problem. In this approach, G is written as

$$G = \sum_{\beta} A_\beta(x, x', z') f_\beta(z), \tag{4.103}$$

where $f_\beta(z)$ are the eigenfunctions of the bilaterally infinite one-dimensional problem discussed in Section II.2 and A_β are coefficients which are to be determined. For the reduced one-dimensional problem in the z-domain, the eigenfunctions are indexed by the continuous variable ζ, and therefore the formal notation in (4.103) becomes (see (4.67)-(4.70))

$$G = \int_{-\infty}^{\infty} d\zeta \; A_\zeta(x, x', z') f_\zeta(z), \tag{4.104}$$

with

$$f_\zeta(z) = \frac{1}{\sqrt{2\pi}} e^{-j\zeta z}. \tag{4.105}$$

The coefficients A_ζ are decomposed into

$$A_\zeta(x, x', z') = g_x(x, x'; \zeta) f_\zeta^*(z'), \tag{4.106}$$

and the Green's function therefore becomes

$$G = \frac{1}{2\pi} \int_{-\infty}^{\infty} g_x(x, x'; \zeta) e^{-j\zeta(z-z')} \, d\zeta. \tag{4.107}$$

Using the representation of G given by (4.107), (4.89) now gives

$$\begin{aligned} \frac{1}{2\pi} \int_{-\infty}^{\infty} \left[\frac{d^2}{dx^2} + k_0^2 - \zeta^2 \right] g_x(x, x'; \zeta) e^{-j\zeta(z-z')} \, d\zeta \\ = -\delta(x - x')\delta(z - z') \\ = \frac{1}{2\pi} \int_{-\infty}^{\infty} -\delta(x - x') e^{-j\zeta(z-z')} \, d\zeta, \end{aligned} \tag{4.108}$$

where the second equality has been obtained by using (4.66). (4.108) implies that

$$\left(\frac{d^2}{dx^2} + k_0^2 - \zeta^2 \right) g_x(x, x'; \zeta) = -\delta(x - x'). \tag{4.109}$$

The boundary conditions associated with (4.109) are

$$g_x = 0 \quad \text{at} \quad x = 0, a, \tag{4.110}$$

and the corresponding solution of (4.125) is (see (4.16))

$$g_x = \frac{\sin \sqrt{\lambda_x} x_< \sin \sqrt{\lambda_x}(a - x_>)}{\sqrt{\lambda_x} \sin \sqrt{\lambda_x} a}, \tag{4.111}$$

where

$$\lambda_x = k_0^2 - \zeta^2 = \xi^2. \tag{4.112}$$

with ξ denoting the x–domain spectral wavenumber. The Green's function G is now

$$G(x, z; x', z'; k_0) = \frac{1}{2\pi} \int_{-\infty}^{\infty} \frac{\sin(\xi x_<) \sin(\xi(a - x_>))}{\xi \sin(\xi a)} e^{-j\zeta(z-z')}\, d\zeta. \qquad (4.113)$$

(4.113) is an expression for G written in terms of the continuous plane wave eigenfunction (mode) spectrum along z, and emphasizes source-excited traveling waves along x, which synthesize the oscillatory-wave closed form result in (4.111) by multiple reflections between the boundaries at $x = 0, a$. The closed-form expression in (4.111) for the x-domain Green's function can be decomposed [11] so as to exhibit the traveling wave hierarchy explicitly.

The phenomenology associated with the z-domain modal plane wave continuum which propagates, and is reflected, along x is totally different from the phenomenology associated with the x-domain discrete modes which propagate along z. The important modes in the continuum of waves represented in (4.113) are established by constructive interference whereas the unimportant modes are filtered out by destructive interference. Thus, constructive interference serves to localize the spectral contributions around the interference maximum, and the integration interval may be localized accordingly. The mathematical technique which implements this scenario is the method of stationary phase. The localization is most pronounced in the "high frequency" range $k_0 a \gg 1$.

Eigenfunction Expansion in the (x, z)-Domain

The (x, z)-domain eigenfunction expansion for $G(x, z; x', z'; k_0^2)$ is written as

$$G = \sum_{\alpha} \int_{-\infty}^{\infty} d\zeta\, g_{\alpha,\zeta}(x', z') F_{\alpha,\zeta}(x, z), \qquad (4.114)$$

where the two-dimensional eigenfunctions $F_{\alpha,\zeta}(x, z)$ are given by (4.85). The expression for G given by (4.114) is now substituted into (4.89), which after using (4.84) and (4.86) gives

$$-\delta(x - x')\delta(z - z') = \sum_{\alpha} \int_{-\infty}^{\infty} d\zeta\, g_{\alpha,\zeta}(x', z') F_{\alpha,\zeta}(x, z)(k_0^2 - \lambda_{\alpha,\zeta}). \qquad (4.115)$$

Using (4.87), (4.115) becomes

$$-\sum_{\alpha} \int_{-\infty}^{\infty} d\zeta\, F_{\alpha,\zeta}(x, z) F_{\alpha,\zeta}^*(x', z') = \sum_{\alpha} \int_{-\infty}^{\infty} d\zeta\, g_{\alpha,\zeta}(x', z') F_{\alpha,\zeta}(x, z)(k_0^2 - \lambda_{\alpha,\zeta}). \qquad (4.116)$$

Since the $\{F_{\alpha,\zeta}\}$ form an orthogonal set, equality of the coefficients in (4.116) yields

$$g_{\alpha,\zeta}(x', z') = -\frac{F_{\alpha,\zeta}^*(x', z')}{k_0^2 - \lambda_{\alpha,\zeta}}. \qquad (4.117)$$

Thus, from (4.114), (4.85) and (4.86), the two-dimensional Green's function becomes

$$G(x,z;x',z';k_0^2) = -\sum_\alpha \int_{-\infty}^{\infty} d\zeta \, \frac{F_{\alpha,\zeta}(x,z)F^*_{\alpha,\zeta}(x',z')}{k_0^2 - \lambda_{\alpha,\zeta}}$$

$$= -\frac{1}{a\pi}\sum_{\alpha=1}^{\infty}\int_{-\infty}^{\infty} d\zeta \, \frac{\sin(\alpha\pi x/a)\sin(\alpha\pi x'/a)e^{-j\zeta(z-z')}}{k_0^2 - [(\alpha\pi/a)^2+\zeta^2]}. \tag{4.118}$$

Using (4.118) to *construct* the completeness relation for the two-dimensional problem is considerably more involved than for the one-dimensional problems in (4.19) and (4.80) since the resolvent complex parameter $\lambda_{xz} = k_0^2$ spans simultaneously the spectral domains λ_x and λ_z associated with the x and z domains, respectively. This will not be pursued further here.

The complete-domain eigenfunctions, in contrast to the reduced-domain eigenfunctions, are sometimes referred to as *resonant eigenfunctions* (or modes). This designation is associated with completely enclosed domains, for which the eigenfunctions $F_\nu(x,z)$ form a discrete set. When this discrete spectrum replaces the discrete-continuous spectrum in the denominator of the integrand in (4.118), the integrand grows indefinitely at the resonant value $k_0^2 = (\omega/c_0)^2 = \lambda_{\alpha,\beta}$, where ω and c_0 are the frequency and wave speed, respectively, associated with the wave equation discussed in Section V. Extending the "resonant" designation also to open domains characterizes in this manner the entire class of complete-domain eigenfunctions, although for continuous spectra the resonant frequencies are not distinct.

Generalized Representations: Relating the Alternatives

The representation of the two-dimensional Green's function $G(x,z;x',z';k_0)$ given by (4.102) has the form

$$G = \sum_\alpha g_z(z,z';\sqrt{k_0^2-\lambda_\alpha})f_\alpha(x)f^*_\alpha(x'). \tag{4.119}$$

According to the generalized completeness relation in (4.21), the eigenfunction sum operator

$$\sum_\alpha f_\alpha(x)f^*_\alpha(x') \tag{4.120}$$

may be replaced by

$$\frac{1}{2\pi j}\oint_{C_x} g_x(x,x';\lambda_x)\,d\lambda_x, \tag{4.121}$$

where $g_x(x,x';\lambda_x)$ is the one-dimensional Green's function associated with the x-domain problem. Thus, (4.119) is equivalent to

$$G = \frac{1}{2\pi j} \oint_{C_x} g_z(z, z'; \sqrt{k_0^2 - \lambda_x}) g_x(x, x'; \lambda_x) \, d\lambda_x, \tag{4.122}$$

with $g_z(z, z'; \sqrt{k_0^2 - \lambda_x})$ and $g_x(x, x'; \lambda_x)$ given explicitly by

$$g_z(z, z'; \sqrt{k_0^2 - \lambda_x}) = \frac{e^{-j\sqrt{k_0^2 - \lambda_x}|z-z'|}}{2j\sqrt{k_0^2 - \lambda_x}} \tag{4.123}$$

and

$$g_x(x, x'; \lambda_x) = \frac{\sin \sqrt{\lambda_x} x_< \sin \sqrt{\lambda_x}(a - x_>)}{\sqrt{\lambda_x} \sin \sqrt{\lambda_x} a}. \tag{4.124}$$

To justify the operational equivalence of (4.120) and (4.121), the contour C_x in (4.122) must enclose *all* of the singularities of $g_x(x, x'; \lambda_x)$ but *none* of the singularities of $g_z(z, z'; \sqrt{k_0^2 - \lambda_x})$. According to (4.123) and (4.124), $g_x(x, x'; \lambda_x)$ has pole singularities at

$$\lambda_x = \lambda_{x,\alpha} = \left(\frac{\alpha\pi}{a}\right)^2, \quad \alpha = 1, 2, 3, \ldots, \tag{4.125}$$

and $g_z(z, z'; \sqrt{k_0^2 - \lambda_x})$ has a branch point at $\lambda_x = k_0^2$. The contour C_x is shown in Figure 4.7, in which the branch point at $\lambda_x = k_0^2$ and the corresponding spectral branch cut are shown slightly above the $\Re\, \lambda_x$ axis in order to clarify the disposition of contours (note that the mapping $\lambda_z = k_0^2 - \lambda_x$ places the z-domain branch cut in the λ_x-plane as shown in Figure 4.7; this corresponds to a branch cut along the positive real axis in the λ_z-plane as in Figure 4.6).

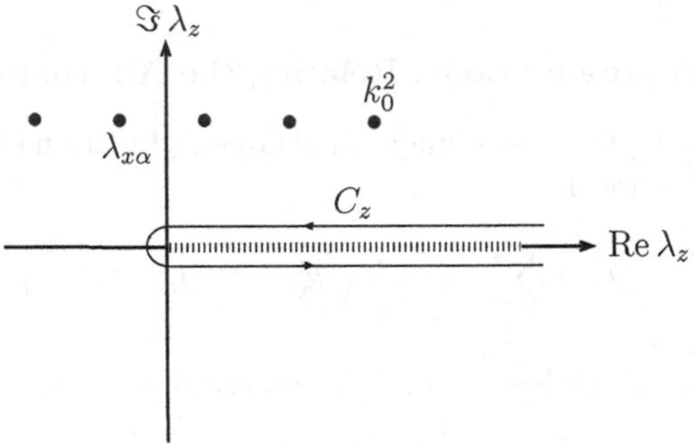

Fig. 4.8. Contours in the λ_z-plane.

Alternatively, the representation of the two-dimensional Green's function $G(x, z; x', z'; k_0)$ given by (4.113) has the form

$$G = \int_{-\infty}^{\infty} d\zeta \; g_x(x, x'; \sqrt{k_0^2 - \zeta^2}) f_\zeta(z) f_\zeta^*(z'). \tag{4.126}$$

As before, the eigenfunction "sum"

$$\int_{-\infty}^{\infty} d\zeta \; f_\zeta(z) f_\zeta^*(z') \tag{4.127}$$

may be replaced by

$$\frac{1}{2\pi j} \oint_{C_z} d\lambda_z \; g_z(z, z'; \lambda_z), \tag{4.128}$$

where $g_z(z, z'; \lambda_z)$ is the one-dimensional Green's function associated with the z-domain problem. Thus, (4.126) is equivalent to

$$G = \frac{1}{2\pi j} \oint_{C_z} g_x(x, x'; \sqrt{k_0^2 - \lambda_z} g_z(z, z'; \lambda_z) \, d\lambda_z. \tag{4.129}$$

The functions $g_x(x, x'; \sqrt{k_0^2 - \lambda_z})$ and $g_z(z, z'; \lambda_z)$ are given by

$$g_x(x, x'; \sqrt{k_0^2 - \lambda_z}) = \frac{\sin \sqrt{k_0^2 - \lambda_z} x_< \sin \sqrt{k_0^2 - \lambda_z}(a - x_>)}{\sqrt{k_0^2 - \lambda_z} \sin \sqrt{k_0^2 - \lambda_z} a} \tag{4.130}$$

and

$$g_z(z, z'; \lambda_z) = \frac{e^{j\sqrt{\lambda_z}|z - z'|}}{2j\sqrt{\lambda_z}}. \tag{4.131}$$

The contour C_z in (4.129) must enclose all of the singularities of $g_z(z, z'; \lambda_z)$ but none of the singularities of $g_x(x, x'; \sqrt{k_0^2 - \lambda_z})$. The function $g_z(z, z'; \lambda_z)$ has a branch point at $\lambda_z = 0$, and the function $g_x(x, x'; \sqrt{k_0^2 - \lambda_z})$ has simple poles at

$$\lambda_z = \lambda_{z,\alpha} = k_0^2 - \left(\frac{\alpha\pi}{a}\right)^2, \quad \alpha = 1, 2, 3, \ldots. \tag{4.132}$$

The contour C_z is shown in Figure 4.8, in which the poles at $\lambda_{z,\alpha}$ are shown slightly above the $\Re\, \lambda_z$ axis for clarity. The two complex plane representations in Figures 4.7 and 4.8 are related via the dispersion relation, i.e.

$$\lambda_z = \sqrt{k_0^2 - \lambda_x^2} = \zeta^2 \tag{4.133}$$

or

$$\lambda_x = \sqrt{k_0^2 - \lambda_z^2} = \xi^2. \tag{4.134}$$

In either the λ_z or the λ_x planes, one representation may be derived from the other by deforming the contours C_z and C_x, respectively, around the singularities of g_x and g_z, respectively. The path deformations can be carried out because of the exponential decay at $|\lambda_{x,z}| \to \infty$ of the synthesizing Green's functions in the integrands.

The (x, z)-domain eigenfunction expansion in (4.118) can be obtained from either the x-domain or the z-domain eigenfunction expansions in (4.102) and (4.113), respectively. In both cases, the respective one-dimensional spectral Green's functions $g_\alpha(z,z')$ or $g_\zeta(x,x')$ are expanded in terms of the z-domain or x-domain eigenfunctions. Thus, referring to (4.101), the z-domain eigenfunction expansion has the form (cf. (2.202))

$$g_\alpha(z,z') = \int_{-\infty}^{\infty} d\zeta \, \frac{f_\zeta(z) f_\zeta^*(z')}{\zeta^2 - (k_0^2 - \lambda_\alpha)}, \tag{4.135}$$

which, upon substitution into (4.102), yields the representation in (4.118). The same result follows from (4.113), except that the roles of $f_\zeta(z)$ and $f_\alpha(x)$ in (4.135) are interchanged.

III Electric Line Source in Radial–Angular Waveguides

III.1 Introduction

In the following part of the chapter we discuss the radial–angular waveguide. We refer again to a two–dimensional domain and naturally many of the techniques and considerations established for rectangular geometries holds also for this case. We will therefore review the main options shortly in order to avoid duplication. We refer to Table 2.10 of Chapter 2 for the range of waveguide geometries accommodated by a coordinate–separable axially–independent $(\rho,\ \phi)$ cylindrical coordinate system. Proceeding in analogy with the parallel–plate waveguide problem in Section II.2 of this chapter, we wish to determine the time–harmonic scalar Green's function (GF) for the Helmholtz equation in cylindrical 2D-$(\rho,\ \phi)$ coordinates,

$$\left(\frac{1}{\rho}\frac{\partial}{\partial \rho}\rho\frac{\partial}{\partial \rho} + \frac{1}{\rho^2}\frac{\partial^2}{\partial \phi^2} + k_0^2\right) G(\rho,\phi;\rho',\phi';k_0) = -\frac{\delta(\rho-\rho')\delta(\phi-\phi')}{\rho} \tag{4.136}$$

with $\rho_1 \le \rho \le \rho_2$ and $\phi_1 \le \phi \le \phi_2$, subject to the PEC boundary conditions,

$$G = 0 \; at \; \rho = \rho_{1,2}; \; \phi = \phi_{1,2} \tag{4.137}$$

at the endpoints $(\rho_{1,2},\ \phi_{1,2})$ of the radial and angular domains, respectively. As in Section II.2 we shall explore the following alternative options based on the reduced 1D problems in the ρ and ϕ coordinates, respectively:

- (1) expressing G in terms of angular (ϕ–domain) eigenfunctions $f_\alpha(\phi)$ and the corresponding radial (ρ–domain) spectral GF, $g_{\rho\alpha}(\rho,\rho')$;
- (2) expressing G in terms of radial eigenfunctions $f_\beta(\rho)$ and the corresponding angular spectral GF, $g_{\phi\beta}(\phi,\phi')$.

III.2 Constituent 1D Problems

Eigenvalue Problem in the ϕ–Domain

Referring to Chapter 2 Section V.3 the angular-domain eigenvalue problem in (2.91),

$$\left(\frac{d^2}{d\phi^2} + \lambda_{\phi\alpha}\right) f_{\phi\alpha}(\phi) = 0 \;\; \lambda_{\phi\alpha} = k_{\phi\alpha}^2 \tag{4.138}$$

is a special case of the generic Sturm–Lioville problem in (2.128), with $p = w = 1, q = 0$, subject to

$$f_{\phi\alpha} = 0 \; at \; \phi = \phi_{1,2}. \tag{4.139}$$

For simplicity (without loss of generality) we set $\phi_1 = 0$, $\phi_2 = \phi_0$. This renders the ϕ–domain problems here identical in form with the x–domain problems in Section II.2:

$$f_{\phi\alpha}(\phi) = \sqrt{\frac{2}{\phi_0}} \sin k_{\phi\alpha}\phi, \;\; k_{\phi\alpha} = \frac{\alpha\pi}{\phi_0} = \sqrt{\lambda_\alpha \phi}. \tag{4.140}$$

From Chapter 2 Section VI.1, it follows furthermore that the eigenfunctions $f_{\phi\alpha}(\phi)$ form the orthogonal set with the completeness relation:

$$\delta(\phi - \phi') = \sum_\alpha f_{\phi\alpha}(\phi) f_{\phi\alpha}^*(\phi'). \tag{4.141}$$

Spectral Green's Function Problem in the ϕ-Domain

By referring to II.2, with the present notation (recalling that $p = w = 1, q = 0$ in (2.128)), one observes that the angular spectral GF defined by

$$\left(\frac{d^2}{d\phi^2} + \lambda_{\phi\alpha}\right) g_\phi(\phi, \phi'; \lambda_\phi) = -\delta(\phi - \phi'), \qquad \lambda_\phi \neq \lambda_{\phi\alpha} \tag{4.142}$$

for $0 \le \phi \le \phi_0$, has the solution (see (4.16)),

$$G_\phi(\phi, \phi'; \lambda_\phi) = \frac{\sin(k_\phi \phi_<) \sin[k_\phi(\phi_0 - \phi_>)]}{k_\phi \sin(k_\phi \phi_0)}, \qquad k_\phi = \sqrt{\lambda_\phi}, \tag{4.143}$$

which exhibits the same convergence behavior as the GF in Section II.2, in the complex λ_ϕ–plane. The corresponding completeness relations in (4.19) and (4.21) become

$$\delta(\phi - \phi') = \frac{1}{2\pi j} \oint_{C_\phi} g_\phi(\phi, \phi'; \lambda_\phi) \, d\lambda_\phi = \sum_\alpha f_{\phi\alpha}(\phi) f_{\phi\alpha}^*(\phi'), \tag{4.144}$$

where C_ϕ denotes a contour which encloses all of the pole singularities of g_ϕ in the complex λ_ϕ–plane. The reduced form in (4.144) agrees with (4.141).

III.3 Eigenvalue Problem in the ρ–Domain

Referring to Section V.3 in Chapter 2, the z–independent version of (2.121) for the 2D (ρ, ϕ) domain implies $k_z \equiv 0$, and the resulting Bessel's equation becomes, in the SL format of (2.128) ($u \to \rho, p \to \rho, q \to \frac{-k_0^2}{\rho}, w \to -\frac{1}{\rho}$),

$$\left(\frac{\partial}{\partial\rho}\rho\frac{\partial}{\partial\rho} + k_0^2\rho - \frac{k_{\rho\beta}^2}{\rho}\right) f_\beta(\rho) = 0, \quad k_{\rho\beta}^2 \equiv \lambda_{\rho\beta} \tag{4.145}$$

for $\rho_1 \leq \rho \leq \rho_2$.

Equation (4.145) is satisfied by a combination of any two linearly independent functions of the form

$$f(\rho) = Z_\tau(k_0\rho)\,, \quad k_\rho = \sqrt{\lambda_\rho}\,, \tag{4.146}$$

where Z_τ stands for any of the following Bessel solutions

$$Z_\tau(\Omega) \to J_\tau(\Omega),\ N_\tau(\Omega),\ H_\tau^{(1)}(\Omega),\ H_\tau^{(2)}(\Omega) \tag{4.147}$$

which represent, respectively, the Bessel, Neumann, and first or second–kind Hankel functions of order τ and argument Ω. Here, the argument is specified as $\Omega = k_0\rho$ and the order $\tau = \bar{\tau} = k_{\rho\beta}$ is the eigenvalue. To satisfy the boundary condition $f_\beta(\rho) = 0$ at $\rho = \rho_{1,2}$, the solution can be constructed as follows:

$$f_\beta(\rho) = A_\beta \left[J_{\bar{\tau}}(k_0\rho_2)N_{\bar{\tau}}(k_0\rho) - N_{\bar{\tau}}(k_0\rho)J_{\bar{\tau}}(k_0\rho_2)\right], \qquad \bar{\tau} = k_{\rho\beta} = \sqrt{\lambda_{\rho\beta}}\,, \tag{4.148}$$

where A_β is an as yet unspecified normalization constant, and the eigenvalues $k_{\rho\beta}$ are determined implicitly via the resonance condition,

$$J_{\bar{\tau}}(k_0\rho_1)N_{\bar{\tau}}(k_0\rho_2) - N_{\bar{\tau}}(k_0\rho_1)J_{\bar{\tau}}(k_0\rho_2) = 0 \ \beta = 1, 2, \ldots. \tag{4.149}$$

Referring again to Sturm-Liouville theory in Section VI.1 in Chapter 2, and recalling the interpretation of (2.128) (preceding (4.145)) for the present problem, yields the orthonormality condition with respect to the weight function $w \to -1/\rho$ (cf. (2.145)),

$$-\int_{\rho_1}^{\rho_2} \frac{1}{\rho} f_\beta(\rho) f_{\hat{\beta}}^*(\rho) d\rho = \delta_{\beta\hat{\beta}}, \tag{4.150}$$

where β and $\hat{\beta}$ are two unequal eigenvalues, and the normalizing constant A_β in (4.148) has been chosen according to (2.144). The completeness relation

$$-\rho\delta(\rho - \rho') = \sum_\beta f_\beta(\rho) f_\beta^*(\rho') \tag{4.151}$$

follows from (2.154).

III.4 Spectral Green's Function Problem in the ρ–Domain

The generic ρ–domain spectral Green's function problem is defined in (2.165) and differs from the eigenvalue problem in that the radial wavenumber k_ρ is a free parameter which can range throughout the complex $k_\rho(=\sqrt{\lambda_\rho})$–plane,

$$\left(\frac{\partial}{\partial\rho}\rho\frac{\partial}{\partial\rho}+k_0^2\rho-\frac{k_{\rho\beta}^2}{\rho}\right)g_\rho(\rho,\rho';\lambda_\rho)=-\delta(\rho-\rho'),\qquad \rho_1\le\rho\le\rho_2 \qquad (4.152)$$

away from the eigenvalues, i.e., $k_\rho\neq k_{\rho\beta}$. The solution in (2.187), within the present format, becomes

$$g_\rho(\rho,\rho')=-\frac{\overleftarrow{f}(\rho_<)\,\overrightarrow{f}(\rho_>)}{W(\overrightarrow{f},\overleftarrow{f})}, \qquad (4.153)$$

where

$$\overleftarrow{f}(\rho)=J_\tau(k_0\rho)N_\tau(k_0\rho_1)-N_\tau(k_0\rho)J_\tau(k_0\rho_1),\quad \tau=k_\rho=\sqrt{\lambda_\rho} \qquad (4.154)$$

and

$$\overrightarrow{f}(\rho)=J_\tau(k_0\rho)N_\tau(k_0\rho_2)-N_\tau(k_0\rho)J_\tau(k_0\rho_2)\,. \qquad (4.155)$$

These wave functions satisfy the homogeneous Bessel equation and the boundary conditions at $\rho=\rho_1$ and $\rho=\rho_2$, respectively. The Wronskian in (2.186) becomes

$$W(\overrightarrow{f},\overleftarrow{f})=\rho\left[\overleftarrow{f}(\rho)\,\overrightarrow{f}'(\rho)-\overrightarrow{f}(\rho)\,\overleftarrow{f}'(\rho)\right], \qquad (4.156)$$

where the prime denoted the ρ–derivative with respect to the argument. The completeness relation in (2.203) becomes

$$-\rho'\delta(\rho-\rho')=\frac{1}{2\pi j}\oint_{C_\rho}g_\rho(\rho,\rho';\lambda_\rho)d\rho=\sum_\beta f_\beta(\rho)f_\beta^*(\rho')\,. \qquad (4.157)$$

In the generalized completeness relation (4.157), C_ρ denotes a contour which encircles all of the pole singularities of g_ρ in the complex λ_ρ plane. Reducing (4.157) through residue evaluation of the contour integral, one obtains (4.151).

III.5 Two–Dimensional Green's Functions: Alternative Representations

Angular Eigenfunctions, Radial Spectral GF

The angular eigenfunction expansion of the Two–dimensional Green's Functions (2DGF) defined in (4.137) takes the form

$$G(\rho, \phi; \rho', \phi'; k_0) = \sum_{\alpha} f_{\phi\alpha}(\phi) f^*_{\phi\alpha}(\phi') g_\rho(\rho, \rho'; \lambda_{\rho\alpha}) \tag{4.158}$$

with $\lambda_{\rho\alpha} = k_0^2 - \frac{\lambda_{\phi\alpha}}{\rho^2}$. To verify the validity of this expansion, substitute (4.158) into the left–hand side of (4.137), interchange the order of summation and differentiation, use (4.138) and (4.142) to eliminate the ϕ–derivatives, use (4.145) and (4.152) to eliminate the ρ–derivatives, and use (4.140) to obtain the expression on the right–hand side of (4.137). The format in (4.158) implies guided propagation of the α–indexed angular modes along the ρ–domain radial waveguide.

Radial Eigenfunctions, Angular Spectral GF

The radial eigenfunction expansion of the 2DGF in (4.137) takes the form

$$G(\rho, \phi; \rho', \phi'; k_0) = \sum_{\beta} f_\beta(\rho) f_\beta(\rho') g_\phi(\phi, \phi'; \lambda_{\phi\beta}). \tag{4.159}$$

Verification of this expansion can be performed by following the analogous sequence of steps described in Section III.4. The format in (4.159) implies propagation of the β–indexed radial modes along the ϕ–domain angular waveguide.

References

[1] R. F. Harrington, *Time Harmonic Electromagnetic Fields.* New York: McGraw-Hill, 1961.

[2] R. E. Collin, *Field Theory of Guided Waves.* New York: IEEE Press, 1991.

[3] J. A. Kong, *Electromagnetic Wave Theory.* Singapore: John Wiley & Sons, 1986.

[4] T. Rozzi and M. Mongiardo, *Open Electromagnetic Waveguides.* London: IEE, 1997.

[5] B. Friedman, *Principles and Techniques of Applied Mathematics.* New York: John Wiley & Sons, 1956.

[6] ——, *Lectures on Applications-Oriented Mathematics.* New York: John Wiley & Sons, 1969.

[7] D. G. Dudley, *Mathematic Foundations for Electromagnetic Theory.* New York: IEEE Press, 1994.

[8] J. R. Wait, *Electromagnetic Wave Theory.* Singapore: John Wiley & Sons, 1987.

[9] ——, *Waves Propagation Theory.* New York: Pergamon Press, 1981.

[10] ——, *Electromagnetic Waves in Stratified Media.* New York: Pergamon Press, 1970.

[11] L. B. Felsen and N. Marcuvitz, *Radiation and Scattering of Waves.* Englewood Cliffs, NJ: Prentice Hall, 1973, Piscataway, NJ: IEEE Press (classic reissue), 1994.

5

Network Representation of Electromagnetic Fields

I Introduction

In the previous chapters we have introduced Maxwell's equation and relevant representations of the Green's functions. The purpose of this chapter is to establish the transition to *numerical field computations* and to introduce the various possibilities arising for network representations.
First we need to pass from functional relationships to their discretized form. To this end, as customarily, we apply the *moment method discretization*, which is briefly recalled in Section II. Then we move to central part of this book, i.e. the rigorous representation of field problems in terms of networks.
As we have discussed in Chapter 1, complex electromagnetic structures may be decomposed into substructures by separating the corresponding spatial domain into *subdomains* joined by common surfaces which represent the *connection network*. Comparing a distributed circuit representing an electromagnetic structure with a lumped element circuit represented by a network, the spatial subdomains may be considered as the circuit elements whereas the complete set of boundary surfaces separating the subdomains corresponds to the connection circuit. Each subdomain, either of finite or infinite extent, may be rigorously characterized by networks.
For a *systematic approach* to electromagnetic field computations in complex structures we divide the geometrical domain into subdomains connected via interfaces. In this way, the task of electromagnetic field computations is separated essentially into:

- Characterization of individual subdomains
- Description of the topology, i.e. of how the subdomains are connected to one other
- Solution of the relative network equations

The problems arising at a *connection surface* have received attention in the literature: In [1] it has been shown that proper care has to be used in order to avoid relative convergence phenomena when selecting the modal basis at the two sides of a step discontinuity. In the context of the mode-matching technique some important properties of waveguide junction generalized scattering matrices have been discussed in [2] and are confirmed by the present approach. Finally, in [3], it was realized that the voltages and currents expressing the amplitudes of the transverse components of the electric and magnetic fields at the interface discontinuity have to satisfy Tellegen's theorem and properties of the normalized generalized scattering matrix were stated [4, 5].

In the present chapter, in Section III, we show that the *connection network*, i.e. the network representation of the transverse field continuity at a connection interface, does not admit an immittance representation, since it does not store any energy. In addition, in Section III.1, we provide criteria for choosing primary and secondary fields at an interface. Finally, in this section, we also introduce *canonical representations* for the connection network.

The connection networks establish the topology and connect various subdomains. The latter can be either of finite extent or of infinite extent. The two cases deserve separate discussion. In Sect. IV we introduce the network representations available for closed regions, i.e. regions of finite volume. We first consider the general case of a certain volume bounded by a surface. The field inside of this volume can be expressed in terms of the resonant modes, i.e. of the three–dimensional vector eigenfunctions. This resonant mode expansion leads, after discretization, to canonical Foster representations and relative network representations. As a more specific case, the finite volume region can exhibit a particular symmetry that suggests the use of a propagating Green's function in one dimension and an eigenfunction expansion in the other two dimensions. Naturally in this case the network representation along the propagation direction is described in terms of transmission lines. It is therefore noted that the theory of alternative Green's function representation provides also alternative network representations. Naturally, a transmission line can be expanded in terms of circuit elements which is also discussed in this section. It is therefore apparent that, in the case of regions (subdomains) of finite volume we have always at least one possibility of deriving the network representation (via the resonant mode expansion) and, when symmetries are present, we can also establish several different networks representations. In the next Section, Sect. V we consider regions that extend up to infinity and, as such, are of infinite volume. For these regions it is not possible to introduce a resonant mode expansion. But, by using radial transmission lines, it is possible to establish rigorous network representations. For example, for objects in free–space we can think of a spherical surface containing these objects. We can then perform a field expansion on the spherical surface in terms of the eigenfunctions corresponding to the finite angular domains (discrete sums). The spherical transmission lines for each spherical mode expansion will now represent propagation in free–space. Naturally, Cauer expansion of the spherical transmission line provides the network representation in terms of circuit elements.

The previous separation of a general field problem into different regions and connecting surfaces and the associated network representations allows systematic solution of field problems. A possible way for such systematic solution is described in Sect. VI. While several other methods of solution are possible the Tableau methods resembles what is done in circuit analysis thus making the analogy between field and circuit problems even more effective.

One of the advantages of the proposed approach is that it permits *use of different numerical methods* in the various subdomains. As a consequence, at each side of a connection surface between adjacent subdomains, we may need to consider different types of expansions for the electric and magnetic field tangential components. For example, one subdomain can be characterized by using modal techniques, i.e. by considering an eigenfunction basis, while the adjacent subdomain can be described by using an integral equation formulation which employs a pulse expansion as a basis. Use of different basis functions has been considered in the past when interfacing purely numerical methods with entire domain boundary conditions; in [6], a modal absorbing boundary condition has been introduced for TLM whereby the inner domain computation was performed by considering a TLM mesh while modal propagation was implemented in the outer domain (a waveguide section). In [7] a waveguide structure was studied and subdivided into two regions, one region being analyzed by modal techniques and the other characterized by finite differences. In these first attempts the interface problem was solved in a heuristic way, without providing a general and systematic solution. The approach discussed in this book makes it clear that it is possible to systematically derive such hybrid methods and the associated network representations.

II Method of Moments

R. F. Harrington has presented in [8] a unified approach to the numerical treatment of field problems by applying the method of moments (MoM). The use of MoM to discretize electromagnetic fields and the availability of high–speed computer allow to reduce the functional equation formulation of an electromagnetic scattering problem into a matrix equation suitable for computer processing. A possible distinction is also feasible between direct and iterative (indirect) MoM. Refer to [8, pp.1-20], [9, pp.7-36], and [10, pp.1-66], for further interesting readings.

Linear field problems are expressed in operator form as

$$\hat{L}(u)F(u) = S(u) \tag{5.1}$$

where $\hat{L}(u)$ represent a linear operator, $S(u)$ is a known function (or source), and $F(u)$ the unknown field. As an example we may establish an equation of the type $\hat{Z}(\boldsymbol{J}) = \boldsymbol{E}$ with $\hat{Z}$ an impedance operator, $\boldsymbol{J}$ the unknown current and $\boldsymbol{E}$ the known (forcing) electric field; or we may deal with an equation of the type

$\hat{Y}(\boldsymbol{E}) = \boldsymbol{H}$ where $\hat{Y}$ is an admittance operator, $\boldsymbol{E}$ is the unknown electric field and $\boldsymbol{H}$ is the known magnetic field.
In order to numerically solve the above equation it is common practice to make use of the method of moments, described in the classic reissue [11]. Therefore in the following we provide just a minimal description of the method of moments while we suggest the interested reader to refer to the original source.
From now on we use the following definition of the inner product

$$\langle \bar{F}, F \rangle \equiv \int_{u_1}^{u_2} \bar{F}^*(u) w(u) F(u) \, du \,, \tag{5.2}$$

where $\bar{F}(u)$ and $F(u)$ are two functions of u and $w(u)$ is a generic weight function. When the operator $\hat{L}(u)$ becomes the SL of (2.127), then the weight function is that relative to the SL operator. Note that, with respect to (2.130), in (5.2) the complex conjugate has been used. From now on the u dependence is not explicitly written in the inner products.
Let us now consider (5.1) and apply the inner product of this equation with testing functions w_α. In the context of the method of moments (MoM) approach the functions w_α are often referred to also as weighting functions. By taking N measurements we obtain the N equations

$$\langle w_\alpha, \hat{L}F \rangle = \langle w_\alpha, S \rangle, \quad \alpha = 1, 2, \ldots, N. \tag{5.3}$$

Let us now look for an approximation of the function F as a linear combination of suitably selected basis functions, or expansion functions, F_β, with unknown amplitude coefficients A_β

$$F(u) = \sum_\beta A_\beta F_\beta(u) \,. \tag{5.4}$$

In the above formula it is common practice to use the same number of basis functions as the number of measurements, although the problem may be solved by a least squares approach when different numbers of basis and testing functions are selected. When selecting as testing and expansion sets the same basis the so-called Galerkin method, which is discussed next in Section II.1, is obtained. By inserting (5.4) into (5.3) yields

$$\sum_\beta A_\beta \langle w_\alpha, \hat{L}F_\beta \rangle = \langle w_\alpha, S \rangle, \quad \alpha = 1, 2, \ldots, N \tag{5.5}$$

By introducing the matrix elements

$$L_{\alpha\beta} = \langle w_\alpha, \hat{L}F_\beta \rangle \tag{5.6}$$

of the linear operator $\hat{L}$ and the expansion coefficients

$$S_\alpha = \langle w_\alpha, S \rangle \tag{5.7}$$

of the function S yields the linear system of equations

$$\sum_\beta L_{\alpha\beta} A_\beta = S_\alpha \tag{5.8}$$

for the determination of the unknown expansion coefficients A_β of the function $F(u)$. Truncating the series expansions with $\alpha = 1...N$ and $\beta = 1...N$ yields a finite-dimensional linear system of equations. With the vectors

$$\mathbf{S} = [S_1, \cdots, S_N]^T \tag{5.9a}$$

$$\mathbf{A} = [A_1, \cdots, A_N]^T \tag{5.9b}$$

and the matrix

$$\widetilde{\mathbf{L}} = \begin{bmatrix} L_{11} & \cdots & L_{1N} \\ \vdots & \ddots & \vdots \\ L_{N1} & \cdots & L_{NN} \end{bmatrix} \tag{5.10}$$

we obtain the linear system of equations in matrix notation

$$\widetilde{\mathbf{L}}\mathbf{A} = \mathbf{S}\,. \tag{5.11}$$

The solution of the linear system of equations (5.11) yields

$$\mathbf{A} = \widetilde{\mathbf{L}}^{-1}\mathbf{S}. \tag{5.12}$$

II.1 Expansion Set

The functions constituting the expansion should be complete, i.e. they should be able to reconstruct whatever type of function. In particular completeness is synthetically expressed as the ability of representing a delta function.
Another important point is that no member of the expansion set should be in the null space of the operator $\hat{L}$. In fact if a function $\bar{F}$ is such that $\hat{L}\bar{F} = 0$, this function $\bar{F}$ can be added with arbitrary amplitude to a solution F, hence making the solution not unique.
Expansion sets may be made by entire–domain basis functions and sub–domain basis. Both choices have some advantages and disadvantages.

Subsectional Basis Functions

We consider two of the simplest class of subsectional basis functions, namely the Dirac delta function and the family of piecewise polynomial interpolation (or spline interpolation).

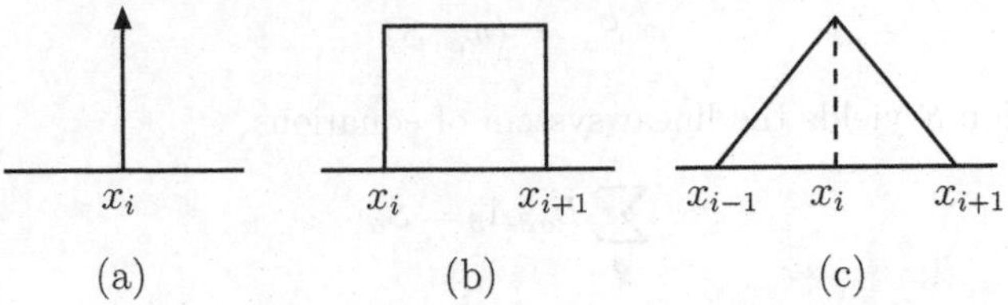

Fig. 5.1. Examples of subsectional basis functions.

Dirac Delta Function

In order to avoid difficult evaluation in integration, the Dirac delta function (see Figure 5.1(a)) is sometimes employed as testing function. Such a procedure is known as point matching

$$B_0 = \delta(x - x_i)\,. \tag{5.13}$$

Splines

The members of the B-spline (bell-spline) family can be generated by the convolution integral,

$$\begin{aligned} B_n(x) &= B_{n-1}(x) * \frac{1}{\Delta}B_1(x) \\ &= \frac{1}{\Delta}\int_{-\frac{\Delta}{2}}^{\frac{\Delta}{2}} B_{n-1}(x - x')dx'\,, \end{aligned} \tag{5.14}$$

where Δ is the size of subsection. The first member, i.e. B-spline of degree 1 is the pulse function (see Figure 5.1(b)),

$$B_1(x) = \begin{cases} 1 & \text{for} \quad x_i < x < x_{i+1}\,, \\ 0 & \text{otherwise} \end{cases} \tag{5.15}$$

while the second member is the triangle function (see Figure 5.1(c)),

$$B_2(x) = \begin{cases} \dfrac{x - x_{i-1}}{x_i - x_{i-1}} & x_{i-1} < x < x_i \\ \dfrac{x_{i+1} - x}{x_{i+1} - x_i} & x_i < x < x_{i+1} \\ 0 & \text{otherwise} \end{cases}\,. \tag{5.16}$$

Pulse functions have limited support and are orthogonal to each other; a pulse functions expansion produces a piecewise-constant representation. Triangle functions are not orthogonal since they overlap between two adjacent subsections, sharing each subsection with the adjacent triangle functions. By superimposing triangle functions along the entire domain of interest, a global piecewise-linear approximation is achieved. However, they can only ensure continuity of the function they represent, but not of the derivative. If continuity of higher derivatives is required, splines of greater degrees and other interpolation polynomials such as the Lagrangian or Hermitian polynomials have to be used [12, pp.192-196], [13, pp.327-368].

The use of wavelet functions, e.g. Haar, Battle-Lemarie, Daubechies wavelets etc. as basis functions has been implemented with certain amount of success in recent years. Because of their oscillatory nature and orthogonal (or biorthogonal) properties, wavelets produce sparse matrices which may offer computational advantages.

Entire-Domain Basis Functions

Well-known examples of complete and orthogonal entire-domain basis functions are the eigenfunctions of the Helmholtz equation for a given domain. As an example the entire-domain functions most suitable for describing the fields in a waveguide are in fact the transverse electric and transverse magnetic modes which compose of harmonic (or sinusoidal) functions. Similarly, suitable functions for describing fields in free space are the spherical modes which are made up of spherical Bessel's functions and spherical harmonics. In bounded or periodic regions, entire-domain basis functions especially eigenfunctions are generally preferred for the sake of fast convergence.

Eigenfunction Expansion

The eigenfunction expansion set satisfies the following equation:

$$\hat{L} f_\alpha = \lambda_\alpha f_\alpha \tag{5.17}$$

with f_α and λ_α denoting eigenfunctions and eigenvalues, respectively. By expanding the unknown function F in terms of the eigenfunctions we get

$$F = \sum_\beta A_\beta f_\beta \tag{5.18}$$

which, after substitution in (5.1),

$$\sum_\beta A_\beta \lambda_\beta f_\beta = S. \tag{5.19}$$

After testing the above equation with the eigenfunction f_α and assuming them orthonormal $\langle f_\beta, f_\alpha \rangle = \delta_{\beta\alpha}$ we have

$$F = \sum_\beta \frac{1}{\lambda_\beta} \langle f_\beta, S \rangle f_\beta . \tag{5.20}$$

It is probably already apparent from the above brief description that the eigenfunction expansion, also called spectral expansion, provides a significant insight in the field problem solution.

Galerkin's Method

Of special mention here is the Galerkin's method which employs identical basis and testing functions, i.e. $F_\beta = w_\beta$. Existing numerical MoM codes based on Galerkin's method yields numerical results more accurate and with better convergence than other choices of basis and testing functions as mentioned in various references as [14, pp.33-36] and [10, pp.40-48].
As shown in [10, pp.41-46], the prevalent choice of Galerkin's method can be attributed to its following properties:

- When the inner product is used for evaluation of the matrix elements, energy is conserved in the approximated solution and the method is in fact equivalent to Rayleigh-Ritz variational method.
- When the symmetric product is used, reciprocity is preserved in the approximation.
- When real-valued basis functions are used, both reciprocity and conservation of energy are preserved in the approximation.
- Since the basis and testing functions are identical and obviously in the same spatial domain, one can circumvent the problem of source singularity inherent in MoM integral equations by exchanging the integral and differential operations.

One should also take note that Galerkin's method will converge to the exact solution for the continuous operator equation if the basis functions are orthogonal and complete in representing F and S over the same spatial domain. It does not necessarily lead to a zero-residual solution, if any of the mentioned properties of the basis functions is not met.

III Regions of Zero Volume: the Connection Network

With reference to Figure 5.2, the boundary separating region $\mathcal{R}_\ell$ from region $\mathcal{R}_k$ is a *connection interface*, i.e. a region of zero volume; it has two sides $\mathcal{B}_{\ell k}$ and $\mathcal{B}_{k\ell}$, to be denoted by Greek letters α, β. By referring directly to the transverse electric and magnetic field components $\boldsymbol{E}_t^\alpha, \boldsymbol{E}_t^\beta$ and $\boldsymbol{H}_t^\alpha, \boldsymbol{H}_t^\beta$, at boundaries α and β, we may express the continuity relationships as

$$\boldsymbol{E}_t^\alpha = \boldsymbol{E}_t^\beta\,, \tag{5.21a}$$

$$\boldsymbol{H}_t^\alpha = \boldsymbol{H}_t^\beta\,. \tag{5.21b}$$

III.1 The Connection Network

Let us consider the fields as expanded on finite orthonormal basis function sets; the assumption of orthonormal basis can be readily removed, if necessary, and is introduced to simplify notation. We consider a set of expansion functions of dimension N_α on side α and a basis of dimension N_β on side β.

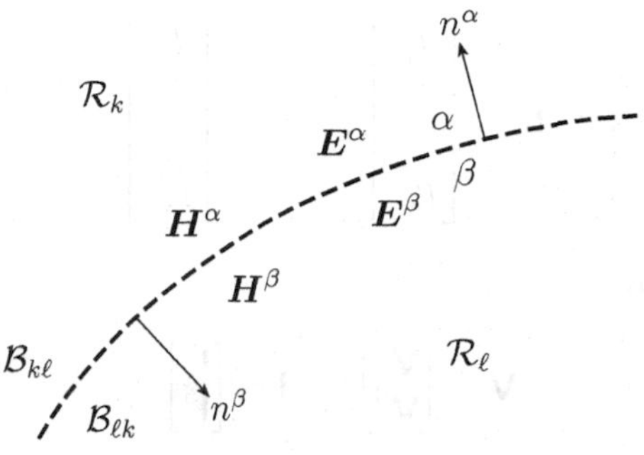

Fig. 5.2. The boundary separating regions $\mathcal{R}_k$ and $\mathcal{R}_\ell$. This boundary can be considered as the surface of a region of zero volume placed between regions $\mathcal{R}_k$ and $\mathcal{R}_\ell$.

Expansion of the Transverse Electric and Magnetic Fields

Subject to the above assumption, we may write the transverse field expansions as

$$(\boldsymbol{E}_t)^\alpha = \sum_n^{N_\alpha} V_n^\alpha \boldsymbol{e}_n^\alpha(\boldsymbol{\rho}), \tag{5.22a}$$

$$(\boldsymbol{E}_t)^\beta = \sum_m^{N_\beta} V_m^\beta \boldsymbol{e}_m^\beta(\boldsymbol{\rho}), \tag{5.22b}$$

$$(\boldsymbol{H}_t)^\alpha = \sum_n^{N_\alpha} I_n^\alpha \boldsymbol{h}_n^\alpha(\boldsymbol{\rho}), \tag{5.22c}$$

$$(\boldsymbol{H}_t)^\beta = \sum_m^{N_\beta} I_m^\beta \boldsymbol{h}_m^\beta(\boldsymbol{\rho}), \tag{5.22d}$$

where the fields are approximated by finite expansions. The vector fields $\boldsymbol{e}_n^\xi(\boldsymbol{\rho})$ and $\boldsymbol{h}_n^\xi(\boldsymbol{\rho})$, $\xi = \alpha, \beta$, are the selected basis functions for electric and magnetic fields. Moreover, V_n^ξ and I_n^ξ, $\xi = \alpha, \beta$, denote the field amplitudes of the electric and magnetic fields, respectively. They are conveniently grouped into the following arrays for the expansions coefficients of the electric field (voltages),

$$\mathbf{V}^\alpha = \begin{bmatrix} V_1^\alpha \\ V_2^\alpha \\ \vdots \\ V_{N_\alpha}^\alpha \end{bmatrix}, \quad \mathbf{V}^\beta = \begin{bmatrix} V_1^\beta \\ V_2^\beta \\ \vdots \\ V_{N_\beta}^\beta \end{bmatrix} \tag{5.23}$$

and for the magnetic fields (currents),

$$\mathbf{I}^{\alpha} = \begin{bmatrix} I_1^{\alpha} \\ I_2^{\alpha} \\ \vdots \\ I_{N_\alpha}^{\alpha} \end{bmatrix}, \quad \mathbf{I}^{\beta} = \begin{bmatrix} I_1^{\beta} \\ I_2^{\beta} \\ \vdots \\ I_{N_\beta}^{\beta} \end{bmatrix} \tag{5.24}$$

leading compactly to

$$\mathbf{V} = \begin{bmatrix} \mathbf{V}^{\alpha} \\ \mathbf{V}^{\beta} \end{bmatrix}, \quad \mathbf{I} = \begin{bmatrix} \mathbf{I}^{\alpha} \\ \mathbf{I}^{\beta} \end{bmatrix}. \tag{5.25}$$

III.2 Tellegen's Theorem for Discretized Fields

We start by considering the expression for power (2.58):

$$\begin{aligned} &\int_{\partial\mathcal{V}} \boldsymbol{E}'(\boldsymbol{\rho}, t') \times \boldsymbol{H}''(\boldsymbol{\rho}, t'') \cdot \boldsymbol{n} dA = \\ &\sum_{n}^{N_\alpha} \sum_{m}^{N_\alpha} V_m^{\alpha'}(t') I_n^{\alpha''}(t'') \int_{\partial\mathcal{V}} \boldsymbol{e}_m^{\alpha} \times \boldsymbol{h}_n^{\alpha} \cdot \boldsymbol{n} dA + \\ &\sum_{n}^{N_\beta} \sum_{m}^{N_\beta} V_m^{\beta'}(t') I_n^{\beta''}(t'') \int_{\partial\mathcal{V}} \boldsymbol{e}_m^{\beta} \times \boldsymbol{h}_n^{\beta} \cdot \boldsymbol{n} dA = 0 \,. \end{aligned} \tag{5.26}$$

By introducing the matrix $\boldsymbol{\Lambda}$ with elements

$$\Lambda_{mn}^{\xi} = \int_{\partial\mathcal{V}} \boldsymbol{e}_m^{\xi} \times \boldsymbol{h}_n^{\xi} \cdot \boldsymbol{n} dA, \tag{5.27}$$

with ξ standing for either α or β, the general form of Tellegen's theorem is

$$\mathbf{V}'^{T}(t) \, \widetilde{\boldsymbol{\Lambda}} \, \mathbf{I}''(t) = 0 \,. \tag{5.28}$$

In general it is convenient to consider orthogonal electric and magnetic field expansions; when this is not the case a suitable orthogonalization process can be carried out providing an orthogonalized basis. In that case the Tellegen's theorem takes the standard form

$$\mathbf{V}'^{T}(t) \, \mathbf{I}''(t) = 0 \,. \tag{5.29}$$

III.3 Testing of the Field Continuity Equations

The discretized form of the field continuity equations in (5.21a) and (5.21b) is obtained by introducing suitable weighting functions $\boldsymbol{w}_m^e$ for the electric field and $\boldsymbol{w}_m^h$ for the magnetic field, which, after insertion of (5.22b) and (5.22d) into (5.21a) and (5.21b), provide the following two sets of electric and magnetic field continuity equations,

$$\sum_{n}^{N_\alpha} V_n^\alpha \langle \boldsymbol{e}_n^\alpha, \boldsymbol{w}_m^e \rangle = \sum_{m}^{N_\beta} V_n^\beta \langle \boldsymbol{e}_n^\beta, \boldsymbol{w}_m^e \rangle \tag{5.30a}$$

$$\sum_{n}^{N_\alpha} I_n^\alpha \langle \boldsymbol{h}_n^\alpha, \boldsymbol{w}_m^h \rangle = \sum_{m}^{N_\beta} I_n^\beta \langle \boldsymbol{h}_n^\beta, \boldsymbol{w}_m^h \rangle \tag{5.30b}$$

Introducing the four matrices $\widetilde{\mathbf{A}}, \widetilde{\mathbf{B}}, \widetilde{\mathbf{C}}, \widetilde{\mathbf{D}}$ whose element are defined as

$$\widetilde{A}_{nm} = \langle \boldsymbol{e}_n^\alpha, \boldsymbol{w}_m^e \rangle \,, \tag{5.31a}$$

$$\widetilde{B}_{nm} = \langle \boldsymbol{e}_n^\beta, \boldsymbol{w}_m^e \rangle \,, \tag{5.31b}$$

$$\widetilde{C}_{nm} = \langle \boldsymbol{h}_n^\alpha, \mathbf{w}_m^h \rangle \,, \tag{5.31c}$$

$$\widetilde{D}_{nm} = \langle \boldsymbol{h}_n^\beta, \boldsymbol{w}_m^h \rangle \tag{5.31d}$$

permits writing the discretized electric field continuity equation as the Kirchhoff Voltage Law (KVL) for the connection network

$$\begin{bmatrix} \widetilde{\mathbf{A}} & -\widetilde{\mathbf{B}} \end{bmatrix} \begin{bmatrix} \mathbf{V}^\alpha \\ \mathbf{V}^\beta \end{bmatrix} = 0 \,. \tag{5.32}$$

Similarly, discretized magnetic field continuity leads to

$$\begin{bmatrix} \widetilde{\mathbf{C}} & -\widetilde{\mathbf{D}} \end{bmatrix} \begin{bmatrix} \mathbf{I}^\alpha \\ \mathbf{I}^\beta \end{bmatrix} = 0 \,, \tag{5.33}$$

which is the Kirchhoff Current Law (KCL) for the connection network.
It is possible to write the above equations in a more compact form by introducing the matrices $\widetilde{\mathbf{K}}$ and $\widetilde{\mathbf{Q}}$

$$\widetilde{\mathbf{K}} = \begin{bmatrix} \widetilde{\mathbf{A}} & -\widetilde{\mathbf{B}} \end{bmatrix} \,, \tag{5.34a}$$

$$\widetilde{\mathbf{Q}} = \begin{bmatrix} \widetilde{\mathbf{C}} & -\widetilde{\mathbf{D}} \end{bmatrix} \,, \tag{5.34b}$$

yielding for the KVL and KCL, respectively,

$$\widetilde{\mathbf{K}}\mathbf{V} = 0 \,, \tag{5.35a}$$

$$\widetilde{\mathbf{Q}}\mathbf{I} = 0 \,. \tag{5.35b}$$

III.4 Independent Quantities

As is well known for networks, not all currents and voltages can be considered as independent; the dimensionality of $\mathbf{V}, \mathbf{I}$ being $N_\alpha + N_\beta$, we can choose N_V independent voltages and N_I independent currents as long as $N_V + N_I = N_\alpha + N_\beta$.

The voltages $\mathbf{V}$ and currents $\mathbf{I}$ relate to the independent voltages $\mathbf{V}^i$ and the independent currents $\mathbf{I}^i$ via [15]

$$\mathbf{I} = \widetilde{\mathbf{K}}^T \mathbf{I}^i, \tag{5.36a}$$

$$\mathbf{V} = \widetilde{\mathbf{Q}}^T \mathbf{V}^i. \tag{5.36b}$$

III.5 Tellegen's Theorem and its Implications

Tellegen's theorem for the connection network can be expressed either as:

$$\mathbf{V}^{T'} \mathbf{I}'' = 0 \tag{5.37}$$

or, by using (5.36b) and (5.36a), in the alternative form

$$\widetilde{\mathbf{Q}} \widetilde{\mathbf{K}}^T = 0. \tag{5.38}$$

This latter expression implies that

$$\widetilde{\mathbf{C}} \widetilde{\mathbf{A}}^T = \widetilde{\mathbf{D}} \widetilde{\mathbf{B}}^T, \tag{5.39}$$

which states the constraints on the matrices $\widetilde{\mathbf{A}}, \widetilde{\mathbf{B}}, \widetilde{\mathbf{C}}, \widetilde{\mathbf{D}}$ in order to satisfy Tellegen's theorem. This means that the weighting functions cannot be chosen arbitrarily but must satisfy the following equation for the generic pair of indices n, k,

$$\sum_m^{N_\alpha} \langle \boldsymbol{w}_n^h, \boldsymbol{h}_m^\alpha \rangle \langle \boldsymbol{e}_m^\alpha, \boldsymbol{w}_k^e \rangle = \sum_m^{N_\beta} \langle \boldsymbol{w}_n^h, \boldsymbol{h}_m^\beta \rangle \langle \boldsymbol{e}_m^\beta, \boldsymbol{w}_k^e \rangle . \tag{5.40}$$

Application of Tellegen's theorem to the connection network thus yields the following result: the weighting functions for testing the field continuity equations have to be selected in accordance with (5.40) in order to provide a consistent set of equations. We shall consider applications of this theorem in some practically relevant cases.

III.6 Application to Orthonormal Bases

The orthonormality condition is expressed as

$$\int_S \mathbf{e}_\ell^\xi(\boldsymbol{\rho}) \cdot \mathbf{e}_k^\xi(\boldsymbol{\rho}) dS = \delta_{\ell k} \tag{5.41}$$

with $\xi = \alpha$ or β, S being the common boundary pertaining to the two expansion sets and $\delta_{\ell k}$ being the Kronecker symbol. It is possible to show that, for orthonormal bases, the Tellegen theorem is always satisfied if we test the electric

field continuity with the electric field basis on one side and test the magnetic field continuity with the magnetic field basis on the other side.
By introducing the coupling matrix $\widetilde{\mathbf{M}}$ whose elements are defined by

$$M_{nm} = \int_S \mathbf{e}_n^\alpha(\boldsymbol{\rho}) \cdot \mathbf{e}_m^\beta(\boldsymbol{\rho}) dS, \tag{5.42}$$

and the identity matrix $\widetilde{\mathbf{1}}$, the following identities hold,

$$\widetilde{\mathbf{M}}\widetilde{\mathbf{M}}^T = \widetilde{\mathbf{1}}, \tag{5.43a}$$

$$\widetilde{\mathbf{M}}^T\widetilde{\mathbf{M}} = \widetilde{\mathbf{1}} \tag{5.43b}$$

thereby satisfying power conservation across the interface.
By using the orthogonality in (5.41) the relationship between voltages and currents is provided in matrix form by the following expressions

$$\mathbf{V}^\alpha = \widetilde{\mathbf{M}}\mathbf{V}^\beta, \tag{5.44a}$$

$$\mathbf{V}^\beta = \widetilde{\mathbf{M}}^T\mathbf{V}^\alpha, \tag{5.44b}$$

$$\mathbf{I}^\alpha = -\widetilde{\mathbf{M}}\mathbf{I}^\beta, \tag{5.44c}$$

$$\mathbf{I}^\beta = -\widetilde{\mathbf{M}}^T\mathbf{I}^\alpha. \tag{5.44d}$$

Equations (5.44a) and (5.44b) represent voltage-controlled voltage sources while (5.44c) and (5.44d) represent current-controlled current sources.
Note that the continuity of the electric field is expressed either by (5.44a) or (5.44b), and the continuity of the magnetic field is expressed either by (5.44c) or (5.44d). Therefore we need to select, in a consistent manner, two out of the four equations (5.44a-5.44d). The Tellegen theorem provides the tool for a consistent choice of the two representative equations as described next.

III.7 Canonical Forms of the Connection Network

Choices of primary and secondary fields which do not violate Tellegen's theorem are either (5.44a) and (5.44d) where $\mathbf{V}^\beta$ and $\mathbf{I}^\alpha$ are the primary network quantities and $\mathbf{V}^\alpha$, $\mathbf{I}^\beta$ are secondary network quantities or (5.44b) and (5.44c) for which the converse holds. According to these choices we may draw the networks shown in Figures 5.3 and 5.4 respectively, based only on ideal transformers, which satisfy (5.44b),(5.44c) and (5.44a),(5.44d).

Properties of the Canonical Connection Networks

It is apparent from the canonical network representations that the scattering matrix is symmetric, $\widetilde{\mathbf{S}}^T = \widetilde{\mathbf{S}}$, orthogonal, $\widetilde{\mathbf{S}}^T\widetilde{\mathbf{S}} = \widetilde{\mathbf{1}}$ and unitary, $\widetilde{\mathbf{S}}\widetilde{\mathbf{S}}^\dagger = \widetilde{\mathbf{1}}$, where the † denotes the Hermitian conjugate matrix.

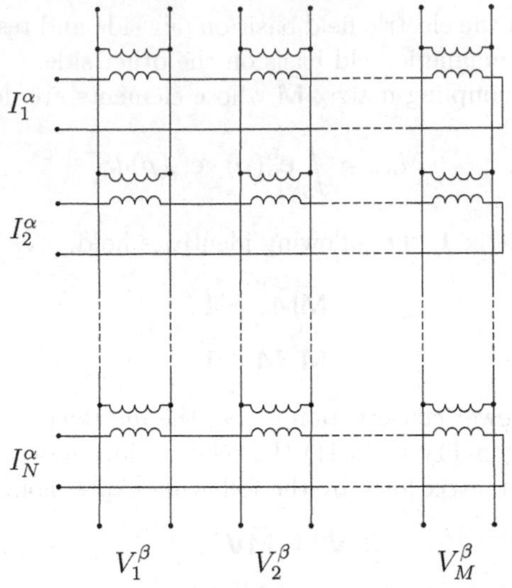

Fig. 5.3. Canonical form of the connection network when using the primary field vectors $\mathbf{V}^\alpha$ (dimension N_α) and $\mathbf{I}^\beta$ (dimension N_β). In this case the secondary fields are $\mathbf{V}^\beta$ (dimension N_β) and $\mathbf{I}^\alpha$ (dimension N_α). In all cases we have $N_\alpha + N_\beta$ primary quantities and the same number of secondary quantities. Scattering representations are also allowed and the connection network is frequency independent.

More Complex Boundaries

We refer now to the bifurcation shown in Figure 5.5 where three different subdomains are joined together. In particular, there is an interface which connects subdomain 1 to subdomain 3, and an interface connecting subdomain 2 to subdomain 3. For brevity, we assume that the electric (magnetic) fields at the interfaces are expanded in terms of suitable basis functions and we express by $\mathbf{V}^i$ ($\mathbf{I}^i$) the vector containing the electric (magnetic) field expansion coefficients relative to region i.

By the same reasoning as in the previous section the connection network for this interface can be obtained by taking $\mathbf{V}^1, \mathbf{V}^2$ and $\mathbf{I}^3$ as independent variables leading to the canonical network representation in the left side of Figure 5.6. The other choice of independent variables is $\mathbf{I}^1, \mathbf{I}^2$ and $\mathbf{V}^3$ which leads to the canonical network shown in the right side of Figure 5.6. Both representations are equally valid to describe the connection network for a bifurcation. Let us now pass from a bifurcation to a step discontinuity.

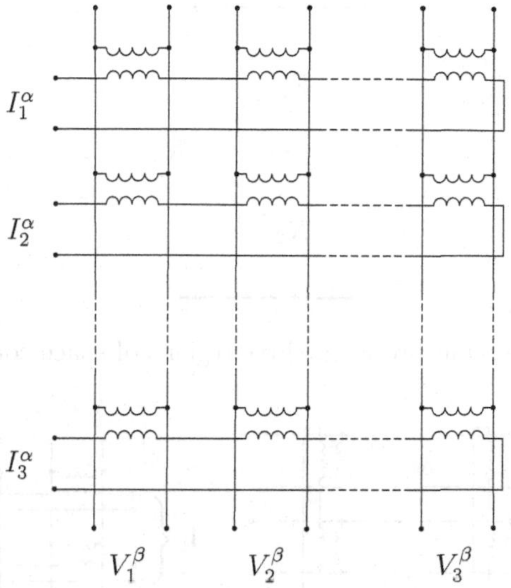

Fig. 5.4. Canonical form of the connection network when using primary field vectors $\mathbf{V}^\beta$ (dimension N_β) and $\mathbf{I}^\alpha$ (dimension N_α). In this case the secondary fields are $\mathbf{V}^\alpha$ (dimension N_α) and $\mathbf{I}^\beta$ (dimension N_β). In all cases we have $N_\beta + N_\alpha$ primary quantities and the same number of secondary quantities. Scattering representations are also allowed and the connection network is frequency independent.

Connection Network for Region Comprising PEC or PMC

For the step discontinuity, region 1 may be considered filled by a PEC, which has to be represented by short-circuits. Thus we need to impose the condition $\mathbf{V}^1 = 0$. The equivalent network is now the one in Figure 5.6 with the ports pertaining to region 1 short-circuited.
Thus in the case of the step discontinuity, the independent quantities are $\mathbf{V}^2$ and $\mathbf{I}^3$, the dependent quantities $\mathbf{I}^2$ and $\mathbf{V}^3$ being determined by the equations

$$\mathbf{V}^3 = \widetilde{\mathbf{M}}\mathbf{V}^2 , \tag{5.45a}$$

$$\mathbf{I}^2 = \widetilde{\mathbf{M}}^T \mathbf{I}^3 . \tag{5.45b}$$

This result, which has been here obtained by network considerations, confirms the one obtained in [2] by a different approach.

IV Network Representations for Regions of Finite Volume

In regions of finite extent it is possible to express the dyadic Green's functions in such a way that a network representation is recovered. Two main cases are possible:

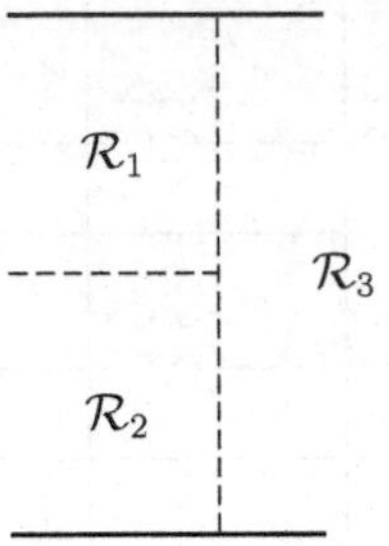

Fig. 5.5. The bifurcation problem: three regions of space connected at an interface.

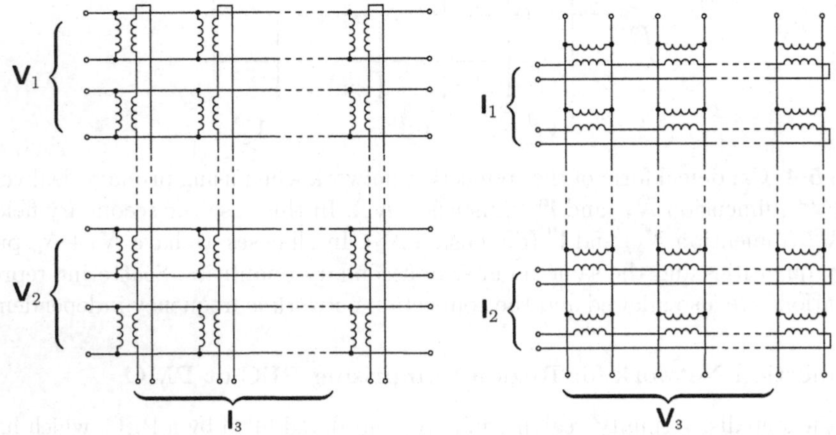

Fig. 5.6. A canonical network for the bifurcation. On the left side $\mathbf{V_1}, \mathbf{V_2}$ and $\mathbf{I_3}$ have been chosen as independent field quantities. On the right side $\mathbf{I_1}, \mathbf{I_2}$ and $\mathbf{V_3}$ have been chosen as independent field quantities.

- The region symmetries suggest a preferred waveguiding direction;
- The region does not present symmetries.

In the first case it is possible to use the dyadic Green's functions introduced in Equations (3.77) and (3.78). The application of the moment procedure to these equations leaves only a transmission line dependence.
In the second case it is necessary to make use of the vector eigenfunction relative to the region under investigation.

IV.1 Foster Representation of the Transmission Line Resonator

It is possible to specify for any linear passive reciprocal circuit an *equivalent Foster* multiport representation. A Foster representation is a *canonical circuit representation* in that sense that it realizes a reactance function with a minimum number of lumped circuit elements. The equivalent Foster admittance multiport representation or Foster impedance representation may be computed analytically from the Green's function. However it is also possible to find an equivalent Foster representation from admittance parameters calculated by numerical field analysis by methods of system identification. Let us consider a lossless transmission line of length l with characteristic impedance Z_0 and propagation speed c, which is short-circuited at one end as depicted in Figure 5.7. This short-circuited transmission line is a reactive oneport. By using standard transmission line analysis we may write for the input impedance Z and for the input admittance Y:

$$Z = jX = jZ_0 \tan \frac{\omega l}{c} , \tag{5.46a}$$

$$Y = jB = -\frac{j}{Z_0} \cot \frac{\omega l}{c} . \tag{5.46b}$$

We introduce the angular frequency

$$\omega_1 = \pi \frac{c}{l} \tag{5.47}$$

which allows to rewrite (5.46a) and (5.46b) as

$$Z = jX = jZ_0 \tan \pi \frac{\omega}{\omega_1} , \tag{5.48a}$$

$$Y = jB = -\frac{j}{Z_0} \cot \pi \frac{\omega}{\omega_1} . \tag{5.48b}$$

Figure 5.8 illustrates the frequency dependence of the reactance X and the susceptance B.

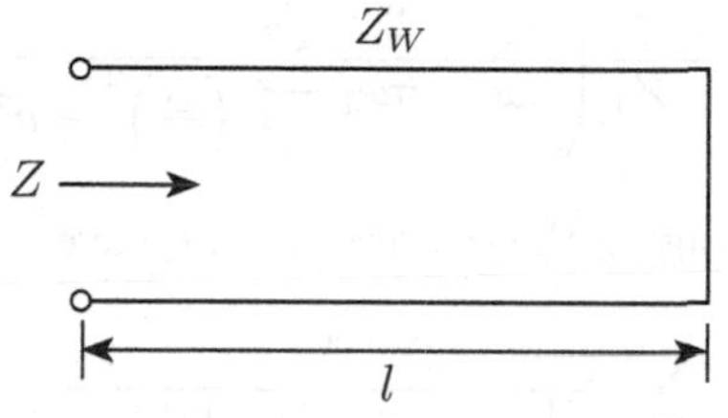

Fig. 5.7. Transmission line of length l short-circuited at one end.

In order to obtain the equivalent circuit for the short-circuited transmission line we perform a *Mittag-Leffler expansion* [16] of $\tan \pi \frac{\omega}{\omega_1}$ and $\cot \pi \frac{\omega}{\omega_1}$ respectively and obtain

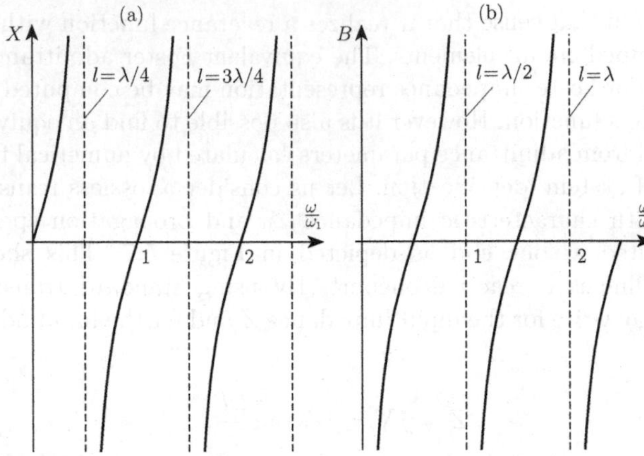

Fig. 5.8. (a) Frequency dependence of the reactance X and (b) the susceptance B of the short-circuited transmission line.

$$\tan \pi x = \frac{2x}{\pi} \sum_{n=1}^{\infty} \frac{1}{\left(n - \frac{1}{2}\right)^2 - x^2}, \tag{5.49a}$$

$$\cot \pi x = \frac{1}{\pi x} + \frac{2x}{\pi} \sum_{n=1}^{\infty} \frac{1}{x^2 - n^2}. \tag{5.49b}$$

After inserting the above expansions into (5.48a) and (5.48b) respectively it follows

$$Z = jZ_0 \frac{2\omega}{\pi \omega_1} \sum_{n=1}^{\infty} \frac{1}{\left(n - \frac{1}{2}\right)^2 - \left(\frac{\omega}{\omega_1}\right)^2}, \tag{5.50a}$$

$$Y = -\frac{j}{Z_0} \left(\frac{\omega_1}{\pi \omega} + \frac{2\omega}{\pi \omega_1} \sum_{n=1}^{\infty} \frac{1}{\left(\frac{\omega}{\omega_1}\right)^2 - n^2} \right), \tag{5.50b}$$

and from this, by rearranging the expressions, we have

$$Z = \sum_{n=1}^{\infty} \frac{1}{j \left[\omega \frac{\pi}{2\omega_1 Z_0} - \frac{1}{\omega} \frac{\left(n - \frac{1}{2}\right)^2 \pi \omega_1}{2 Z_0} \right]}, \tag{5.51a}$$

$$Y = \frac{1}{j\omega \frac{\pi Z_0}{\omega_1}} + \sum_{n=1}^{\infty} \frac{1}{j \left[\omega \frac{\pi Z_0}{2\omega_1} - \frac{1}{\omega} \frac{n^2 \pi \omega_1 Z_0}{2} \right]}. \tag{5.51b}$$

If we now introduce the following quantities:

$$C_p = \frac{\pi}{2\omega_1 Z_0}, \tag{5.52a}$$

$$L_0 = \frac{\pi Z_0}{\omega_1}, \quad L_s = \frac{\pi Z_0}{2\omega_1}, \tag{5.52b}$$

$$L_{pn} = \frac{2Z_0}{\left(n - \frac{1}{2}\right)^2 \pi\omega_1}, \tag{5.53a}$$

$$C_{sn} = \frac{2}{n^2 \pi \omega_1 Z_0}, \tag{5.53b}$$

$$\omega_{0pn} = \frac{1}{\sqrt{C_p L_{pn}}} = \left(n - \frac{1}{2}\right)\omega_1, \tag{5.54a}$$

$$\omega_{0sn} = \frac{1}{\sqrt{L_s C_{sn}}} = n\omega_1 \tag{5.54b}$$

we obtain

$$Z = \sum_{n=1}^{\infty} \frac{1}{j\left(\omega C_p - \frac{1}{\omega L_{pn}}\right)}, \tag{5.55a}$$

$$Y = \frac{1}{j\omega L_0} + \sum_{n=1}^{\infty} \frac{1}{j\left(\omega L_s - \frac{1}{\omega C_{sn}}\right)}. \tag{5.55b}$$

These *fractional expansion representations* are called the *Foster representations* [17,18]. The *Foster impedance representation*, also called *Foster representation of the first kind*, given by (5.55a) describes the series connection of an infinite number of parallel resonant circuits with resonance frequencies given by (5.54a), whereas the *Foster admittance representation*, also called *Foster representation of the second kind*, (5.55b) describes the parallel connection of an infinite number of series resonant circuits and one inductance L_0 where the resonant frequencies of the series resonant circuits are given by (5.54b). The corresponding *equivalent circuits* are the *Foster equivalent circuit of the first kind* shown in Figure 5.9a and the the *Foster equivalent circuit of the second kind* shown in Figure 5.9b.
For lossy transmission lines we have to add loss resistors in the equivalent circuits. In the case of Figure 5.9a we have to add a loss conductor in parallel to each parallel resonant circuit, and in the case of Figure 5.9b we have to add a loss resistor in series to each series resonant circuit. Considering a transmission line resonator at frequencies ω_{0pn} or ω_{0sn} in the neighborhood of one pole of the reactance function allows to neglect all poles with the exception of the pole under consideration. By this way the equivalent circuit may be reduced to a single resonant circuit describing the pole under consideration. Figure 5.10 shows the corresponding equivalent circuits consisting of a single parallel or series resonant circuit respectively.

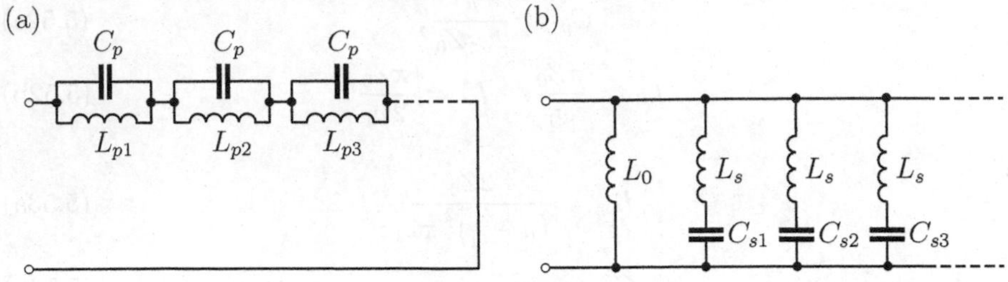

Fig. 5.9. Equivalent circuits of the lossless transmission line resonator (a) according to (5.55a) and (b) according to (5.55b).

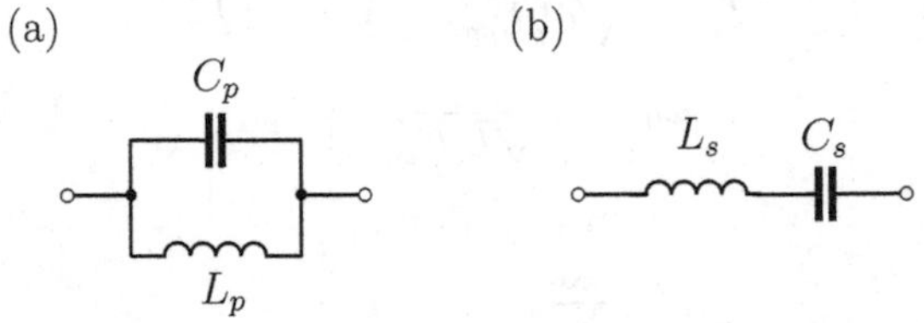

Fig. 5.10. Equivalent circuits of the lossless transmission line resonator (a) near a parallel resonance and (b) near a series resonance.

IV.2 Green's Function and Multiport Foster Representation

We now consider a domain $\mathcal{R}_n$ with the tangential electric and magnetic field components on the boundary $\partial\mathcal{R}_n$ given by $\boldsymbol{E}_t$ and $\boldsymbol{H}_t$. These tangential field components are related via

$$\boldsymbol{E}_t(\boldsymbol{r},\omega) = \int_{\partial\mathcal{R}_n} \mathscr{Z}(\boldsymbol{r},\boldsymbol{r}',\omega)\,\boldsymbol{H}_t(\boldsymbol{r},\omega)\,d\mathcal{A} \tag{5.56}$$

or

$$\boldsymbol{H}_t(\boldsymbol{r},\omega) = \int_{\partial\mathcal{R}_n} \mathscr{Y}(\boldsymbol{r},\boldsymbol{r}',\omega)\,\boldsymbol{E}_t(\boldsymbol{r},\omega)\,d\mathcal{A}\,, \tag{5.57}$$

where $\mathscr{Z}(\boldsymbol{r},\boldsymbol{r}',\omega)$ and $\mathscr{Y}(\boldsymbol{r},\boldsymbol{r}',\omega)$ are the dyadic Green's functions in the impedance representation or admittance representation, respectively. The Green's functions $\mathscr{Z}(\boldsymbol{r},\boldsymbol{r}',\omega)$ and $\mathscr{Y}(\mathbf{x},\boldsymbol{r}',\omega)$ are given by [19]

$$\mathscr{Z}(\boldsymbol{r},\boldsymbol{r}',\omega) = \frac{1}{j\omega}\mathscr{Z}^0(\boldsymbol{r},\boldsymbol{r}') + \sum_\lambda \frac{1}{j\omega}\frac{\omega^2}{\omega^2-\omega_\lambda^2}\mathscr{Z}^\lambda(\boldsymbol{r},\boldsymbol{r}') \tag{5.58}$$

and

$$\mathscr{Y}(\boldsymbol{r},\boldsymbol{r}',\omega) = \frac{1}{j\omega}\mathscr{Y}^0(\boldsymbol{r},\boldsymbol{r}') + \sum_\lambda \frac{1}{j\omega}\frac{\omega^2}{\omega^2-\omega_\lambda^2}\mathscr{Y}^\lambda(\boldsymbol{r},\boldsymbol{r}')\,. \tag{5.59}$$

The dyadics $\mathcal{Z}^0(\boldsymbol{r},\boldsymbol{r}')$ and $\mathcal{Y}^0(\boldsymbol{r},\boldsymbol{r}')$ represent the static parts of the Green's functions, whereas each term $\mathcal{Z}^\lambda(\boldsymbol{r},\boldsymbol{r}')$ and $\mathcal{Y}^\lambda(\boldsymbol{r},\boldsymbol{r}')$, respectively, corresponds to a pole at the frequency ω_λ.
We discretize (5.56) and (5.57) by expanding the tangential fields on $\partial\mathcal{R}_n$ into a complete set of vector orthonormal basis functions. These expansions need only to be valid on $\partial\mathcal{R}_n$,

$$\boldsymbol{E}_t(\boldsymbol{r},\omega) = \sum_n E_n(\omega)\boldsymbol{u}_n(\boldsymbol{r}), \tag{5.60a}$$

$$\boldsymbol{H}_t(\boldsymbol{r},\omega) = \sum_n H_n(\omega)\boldsymbol{v}_n(\boldsymbol{r}). \tag{5.60b}$$

The vector basis functions $\boldsymbol{u}_n(\boldsymbol{r})$ and $\boldsymbol{v}_n(\boldsymbol{r})$ fulfill the orthogonality relations

$$\int_{\partial\mathcal{R}_n} \boldsymbol{u}_m^*(\boldsymbol{r})\cdot\boldsymbol{u}_n(\boldsymbol{r})\,d\mathcal{A} = \delta_{mn}, \tag{5.61a}$$

$$\int_{\partial\mathcal{R}_n} \boldsymbol{v}_m^*(\boldsymbol{r})\cdot\mathbf{v}_n(\boldsymbol{r})\,d\mathcal{A} = \delta_{mn}. \tag{5.61b}$$

Furthermore the two sets of vector basis functions $\boldsymbol{u}_n(\boldsymbol{r})$ and $\boldsymbol{v}_n(\boldsymbol{r})$ are related via

$$\boldsymbol{v}_n(\boldsymbol{r}) = \mathbf{n}(\boldsymbol{r}) \times \boldsymbol{u}_n(\boldsymbol{r}), \tag{5.62a}$$

$$\boldsymbol{u}_n(\boldsymbol{r}) = \mathbf{n}(\boldsymbol{r}) \times \boldsymbol{v}_n(\boldsymbol{r}), \tag{5.62b}$$

where $\mathbf{n}(\boldsymbol{r})$ is the normal vector on $\partial\mathcal{R}_n$. The expansion coefficients E_n and H_n may be considered as generalized voltages and currents. From (5.60a) and (5.60b) and the orthogonality relations (5.61a) and (5.61b) we obtain

$$E_n(\omega) = \int_{\partial\mathcal{R}_n} \boldsymbol{u}_n^*(\boldsymbol{r})\cdot\boldsymbol{E}_t(\boldsymbol{r},\omega)\,d\mathcal{A}, \tag{5.63a}$$

$$H_n(\omega) = \int_{\partial\mathcal{R}_n} \boldsymbol{v}_n^*(\boldsymbol{r})\cdot\boldsymbol{H}_t(\boldsymbol{r},\omega)\,d\mathcal{A}. \tag{5.63b}$$

If the domain $(\mathcal{V})$ is partially bounded by an ideal electric or magnetic wall $\boldsymbol{E}_t$ or $\boldsymbol{H}_t$ respectively vanish on these walls. If the independent field variable vanishes on the boundary, this part of the boundary does not need to be represented by the basis functions. If only electric walls are involved, the admittance representation of the Green's function will be appropriate, and if only magnetic walls are involved, the impedance representation will be appropriate. Let us consider the domain in Figure 1.1. In this case, the main part of the boundary $(\partial\mathcal{V})$ is formed by an electric wall. Only ports 1 and 2 are left open. Choosing the admittance representation, we only need to expand the field on the port surfaces into basis functions. Applying the moment method, we obtain

$$Z_{m,n}(\omega) = \int_{\partial \mathcal{R}_n} \boldsymbol{u}^*_m(\boldsymbol{r}) \cdot \mathscr{Z}(\boldsymbol{r}, \boldsymbol{r}', \omega) \cdot \boldsymbol{v}_n(\boldsymbol{r})\, d\mathcal{A}, \tag{5.64a}$$

$$Y_{m,n}(\omega) = \int_{\partial \mathcal{R}_n} \boldsymbol{v}^*_m(\boldsymbol{r}) \cdot \mathscr{Y}(\boldsymbol{r}, \boldsymbol{r}', \omega) \cdot \boldsymbol{u}_n(\boldsymbol{r})\, d\mathcal{A}. \tag{5.64b}$$

Then from (5.58) and (5.59), the impedance matrix terms $Z_{m,n}(\omega)$ and the admittance matrix terms $Y_{m,n}(\omega)$ may be represented by

$$Z_{m,n}(\omega) = \frac{1}{j\omega} z^0_{mn} + \sum_\lambda \frac{1}{j\omega} \frac{\omega^2}{\omega^2 - \omega_\lambda^2} z^\lambda_{mn}, \tag{5.65a}$$

$$Y_{m,n}(\omega) = \frac{1}{j\omega} y^0_{mn} + \sum_\lambda \frac{1}{j\omega} \frac{\omega^2}{\omega^2 - \omega_\lambda^2} y^\lambda_{mn}. \tag{5.65b}$$

IV.3 The Canonical Foster Representation of Distributed Circuits

For a linear reciprocal lossless multiport an equivalent circuit model may be specified by the canonical Foster representation ([20], pp. 197–199), [17]. Figure 5.11a shows a *compact reactance multiport* describing a pole at the frequency ω_λ. This compact multiport consists of one series resonant circuit and M ideal transformers. The admittance matrix of this compact multiport is given by

$$\widetilde{\mathbf{Y}}_\lambda(\omega) = \frac{1}{j\omega L_\lambda} \frac{\omega^2}{\omega^2 - \omega_\lambda^2} \widetilde{\mathbf{A}}_\lambda \tag{5.66}$$

with the real frequency-independent rank 1 matrix $\widetilde{\mathbf{A}}_\lambda$ given by

$$\widetilde{\mathbf{A}}_\lambda = \begin{bmatrix} n_{\lambda 1} \\ n_{\lambda 2} \\ n_{\lambda 3} \\ \vdots \\ n_{\lambda N} \end{bmatrix} \begin{bmatrix} n_{\lambda 1} \\ n_{\lambda 2} \\ n_{\lambda 3} \\ \vdots \\ n_{\lambda N} \end{bmatrix}^T = \begin{bmatrix} n^2_{\lambda 1} & n_{\lambda 1} n_{\lambda 2} & n_{\lambda 1} n_{\lambda 3} & \cdots & n_{\lambda 1} n_{\lambda N} \\ n_{\lambda 2} n_{\lambda 1} & n^2_{\lambda 2} & n_{\lambda 2} n_{\lambda 3} & \cdots & n_{\lambda 2} n_{\lambda N} \\ n_{\lambda 3} n_{\lambda 1} & n_3 n_{\lambda 2} & n^2_{\lambda 3} & \cdots & n_{\lambda 3} n_{\lambda N} \\ \vdots & \vdots & \vdots & \ddots & \vdots \\ n_{\lambda N} n_{\lambda 1} & n_{\lambda N} n_{\lambda 2} & n_{\lambda N} n_{\lambda 3} & \cdots & n^2_{\lambda N} \end{bmatrix}. \tag{5.67}$$

The $n_{\lambda i}$ are the turns ratios of the ideal transformers in Figure 5.11a. Figure 5.11b shows a *compact reactance multiport* describing a pole at the frequency $\omega = 0$. The admittance matrix of this compact multiport is given by

$$\widetilde{\mathbf{Y}}_0 = \frac{1}{j\omega L_0} \widetilde{\mathbf{A}}_0, \tag{5.68}$$

where $\widetilde{\mathbf{A}}_0$ is a real frequency independent rank 1 matrix as defined in (5.67) If the admittance matrix is of rank higher than 1 it has to be decomposed into a sum of rank 1 matrices. Each rank 1 matrix corresponds to a compact multiport. The complete admittance matrix describing a circuit with a finite number of poles is obtained by parallel connecting the circuits describing the individual poles. In

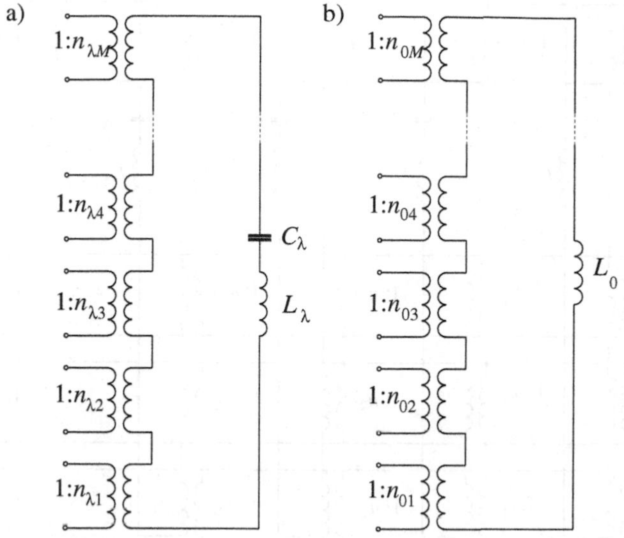

Fig. 5.11. A compact series multiport element representing a pole a) at $\omega = \omega_\lambda$ and b) at $\omega = 0$.

the *Foster admittance representation*, the admittance matrix $\widetilde{\mathbf{Y}}(\omega)$ is given by

$$\widetilde{\mathbf{Y}}(\omega) = \frac{1}{j\omega L_0}\widetilde{\mathbf{A}}_0 + \sum_{\lambda=1}^{N} \frac{1}{j\omega L_\lambda} \frac{\omega^2}{\omega^2 - \omega_\lambda^2} \widetilde{\mathbf{A}}_\lambda . \tag{5.69}$$

This admittance matrix describes a parallel connection of elementary multiports, each of which consists of a series resonant circuit and an ideal transformer. Figure 5.12 shows the complete circuit of the Foster admittance representation. There exists also a dual impedance representation where elementary circuits consisting of parallel resonant circuits and ideal transformers are connected in series. Figure 5.13a shows a *compact reactance multiport* describing a pole at the frequency ω_λ. This compact multiport consists of one parallel circuit and M ideal transformers. The impedance matrix of this compact multiport is given by

$$\widetilde{\mathbf{Z}}_\lambda(\omega) = \frac{1}{j\omega C_\lambda} \frac{\omega^2}{\omega^2 - \omega_\lambda^2} \widetilde{\mathbf{B}}_\lambda \tag{5.70}$$

with the real frequency independent rank 1 matrix $\widetilde{\mathbf{B}}_\lambda$ given by

$$\widetilde{\mathbf{B}}_\lambda = \begin{bmatrix} n_{\lambda 1} \\ n_{\lambda 2} \\ n_{\lambda 3} \\ \vdots \\ n_{\lambda N} \end{bmatrix} \begin{bmatrix} n_{\lambda 1} \\ n_{\lambda 2} \\ n_{\lambda 3} \\ \vdots \\ n_{\lambda N} \end{bmatrix}^T = \begin{bmatrix} n_{\lambda 1}^2 & n_{\lambda 1}n_{\lambda 2} & n_{\lambda 1}n_{\lambda 3} & \cdots & n_{\lambda 1}n_{\lambda N} \\ n_{\lambda 2}n_{\lambda 1} & n_{\lambda 2}^2 & n_{\lambda 2}n_{\lambda 3} & \cdots & n_{\lambda 2}n_{\lambda N} \\ n_{\lambda 3}n_{\lambda 1} & n_3 n_{\lambda 2} & n_{\lambda 3}^2 & \cdots & n_{\lambda 3}n_{\lambda N} \\ \vdots & \vdots & \vdots & \ddots & \vdots \\ n_{\lambda N}n_{\lambda 1} & n_{\lambda N}n_{\lambda 2} & n_{\lambda N}n_{\lambda 3} & \cdots & n_{\lambda N}^2 \end{bmatrix} . \tag{5.71}$$

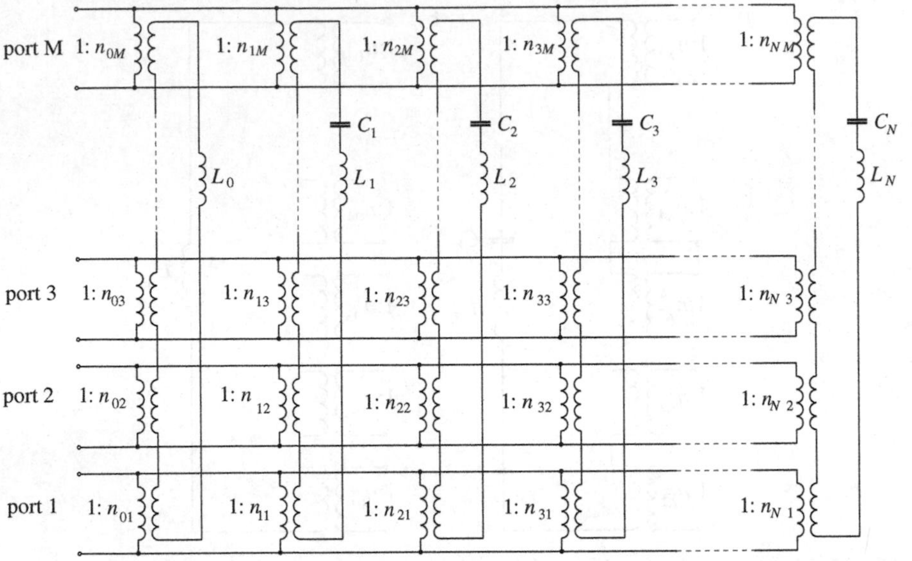

Fig. 5.12. Foster admittance representation of a multiport.

Figure 5.13b shows a *compact reactance multiport* describing a pole at the frequency $\omega = 0$. The impedance matrix of this compact multiport is given by

$$\widetilde{\mathbf{Z}}_0 = \frac{1}{j\omega L_0} \widetilde{\mathbf{B}}_0 , \tag{5.72}$$

where $\widetilde{\mathbf{B}}_0$ is a real frequency independent rank 1 matrix as defined in (5.67). The complete impedance matrix describing a circuit with a finite number of poles is obtained by series connecting the circuits describing the individual poles. In the *Foster impedance representation*, the impedance matrix $\widetilde{\mathbf{Z}}(p)$ is given by

$$\widetilde{\mathbf{Z}}(\omega) = \frac{1}{j\omega C_0} \widetilde{\mathbf{B}}_0 + \sum_{\lambda=1}^{N} \frac{1}{j\omega C_\lambda} \frac{\omega^2}{\omega^2 - \omega_\lambda^2} \widetilde{\mathbf{B}}_\lambda \tag{5.73}$$

Figure 5.14 shows the complete circuit of the Faster impedance representation.

V Regions Extending to Infinity: Radiation Problems

Let us assume the complete electromagnetic structure under consideration embedded in a virtual sphere $\mathcal{S}$ as shown in Figure 5.15. Outside the sphere free space is assumed. The complete electromagnetic field outside the sphere may be expanded into a set of TM and TE spherical waves propagating in outward direction. In 1948 L.J. Chu in his paper on physical limitations of omni–directional antennas has investigated the orthogonal mode expansion of the radiated field [21]. Using

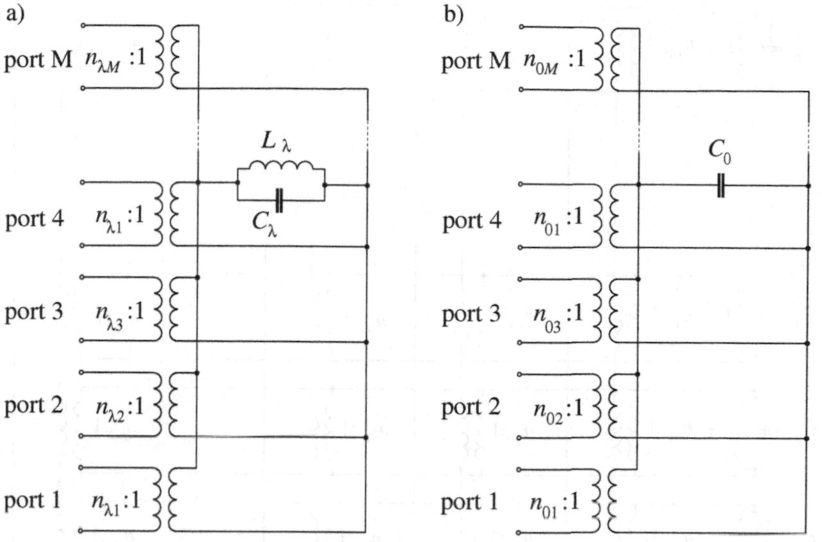

Fig. 5.13. A compact parallel multiport element representing a pole a) at $\omega = \omega_\lambda$ and b) at $\omega = 0$.

the recurrence formula for spherical bessel functions he gave the Cauer representation [17,20] of the equivalent circuits of the TM_n and the TE_n spherical waves. The equivalent circuit expansion of spherical waves also is treated in the books of Harrington [18] and Felsen [22].

The TM modes are given by

$$\boldsymbol{H}_{mn}^{TMij} = \nabla \times \left(A_{mn}^{ij} \boldsymbol{e}_r \right) , \tag{5.74a}$$

$$\boldsymbol{E}_{mn}^{TMij} = \frac{1}{j\omega\varepsilon} \nabla \times \boldsymbol{H}_{mn}^{TMi} , \tag{5.74b}$$

where $n = 1, 2, 3, 4, \ldots, m = 1, 2, 3, 4, \ldots, n$, $i = e, o$, and $j = 1, 2$. The radial component A_{mn}^{ij} of the vector potential is given by

$$A_{mn}^{ej} = a_{mn}^{ej} P_n^m(\cos\theta) \cos m\varphi \, h_n^{(j)}(kr) , \tag{5.75a}$$

$$A_{mn}^{oj} = a_{mn}^{oj} P_n^m(\cos\theta) \sin m\varphi \, h_n^{(j)}(kr) , \tag{5.75b}$$

where the $P_n^m(\cos\theta)$ are the associated Legendre polynomials and $h_n^{(j)}(kr)$ are the spherical Hankel functions. The a_{mn}^{ej} and a_{mn}^{oj} are coefficients. Inward propagating waves are represented by $h_n^{(1)}(kr)$ and outward propagating waves are represented by $h_n^{(2)}(kr)$. Since outside the sphere, for $r > r_0$ no sources exist, only outward propagating waves occur and we have only to consider the spherical Hankel functions $h_n^{(2)}(kr)$.

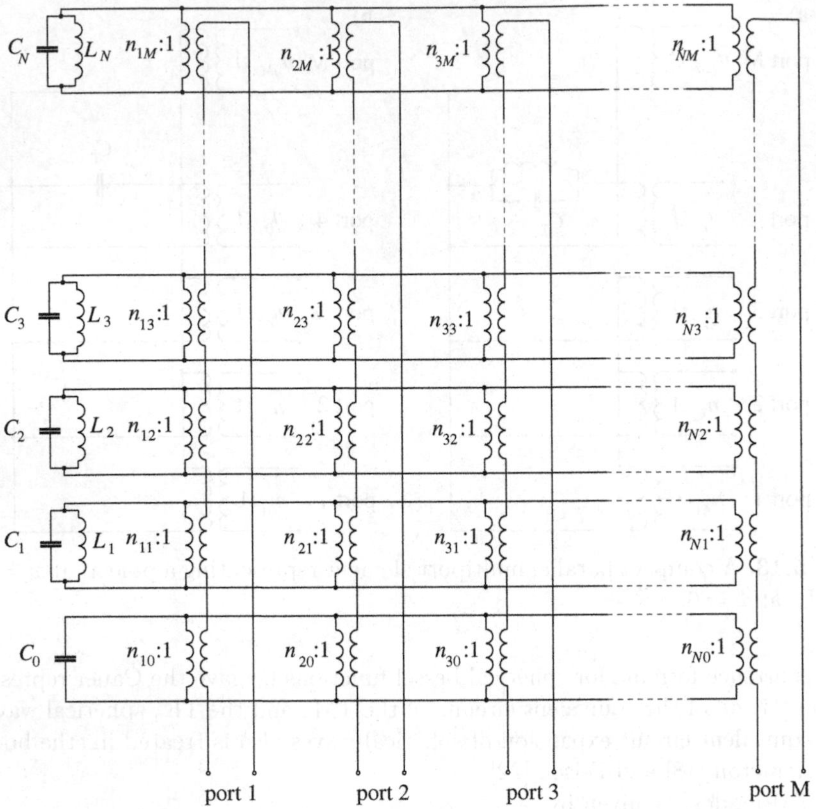

Fig. 5.14. Foster impedance representation of a multiport

The TE modes are dual with respect to the TM modes and are given by

$$\boldsymbol{E}_{mn}^{TEij} = -\nabla \times \left(F_{mn}^{ij} \boldsymbol{e}_r\right), \tag{5.76a}$$

$$\boldsymbol{H}_{mn}^{TEij} = -\frac{1}{j\omega\varepsilon} \nabla \times \boldsymbol{E}_{mn}^{TMi}, \tag{5.76b}$$

where $n = 1, 2, 3, 4, \ldots$, $m = 1, 2, 3, 4, \ldots, n$, $i = e, o$, and $j = 1, 2$. The radial component F_{mn}^{ij} of the dual vector potential is given by

$$F_{mn}^{ej} = f_{mn}^{ej} P_n^m(\cos\theta) \cos m\varphi \, h_n^{(j)}(kr), \tag{5.77a}$$

$$F_{mn}^{oj} = f_{mn}^{oj} P_n^m(\cos\theta) \sin m\varphi \, h_n^{(j)}(kr). \tag{5.77b}$$

where the $P_n^m(\cos\theta)$ are the associated Legendre polynomials and $h_n^{(j)}(kr)$ are the spherical Hankel functions. The f_{mn}^{ej} and f_{mn}^{oj} are coefficients.
The wave impedances for the outward propagating TM and TE modes are given by

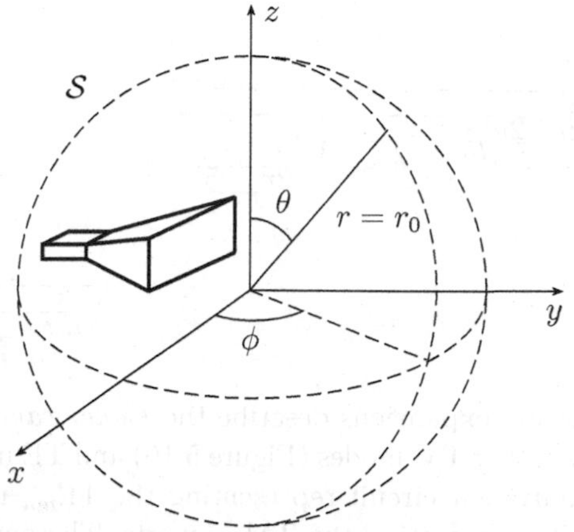

Fig. 5.15. Embedding of an electromagnetic structure into a sphere.

$$Z^{+}_{mn} = \frac{E^{+}_{mn\theta}}{H^{+}_{mn\varphi}} = -\frac{E^{+}_{mn\varphi}}{H^{+}_{mn\theta}}\,. \tag{5.78}$$

The superscript + denotes the outward propagating wave. For the TM and TE modes we obtain

$$Z^{+TM}_{mn} = j\eta \frac{\frac{\mathrm{d}}{\mathrm{d}r}\left(rh_n^{(2)}(kr)\right)}{rh_n^{(2)}(kr)}\,, \tag{5.79a}$$

$$Z^{+TE}_{mn} = -j\eta \frac{rh_n^{(2)}(kr)}{\frac{\mathrm{d}}{\mathrm{d}r}\left(rh_n^{(2)}(kr)\right)}\,, \tag{5.79b}$$

where $\eta = \sqrt{\mu/\varepsilon}$ is the wave impedance of the plane wave. We note that the characteristic wave impedances only depend on the index n and the radius r_0 of the sphere.

V.1 The Cauer Canonic Representation of Radiation Modes

Using the recurrence formulae for Hankel functions we perform continued fraction expansions of the wave impedances of the TM modes

$$Z^{+TM}_{mn} = \eta \left\{ \frac{n}{jkr} + \cfrac{1}{\cfrac{2n-1}{jkr} + \cfrac{1}{\cfrac{2n-3}{jkr} + \ddots + \cfrac{1}{\cfrac{3}{jkr} + \cfrac{1}{\cfrac{1}{jkr} + 1}}}} \right\} \tag{5.80}$$

and the TE modes

$$Z_{mn}^{+TE} = \eta \left\{ \cfrac{1}{\cfrac{n}{jkr} + \cfrac{1}{\cfrac{2n-1}{jkr} + \cfrac{1}{\cfrac{2n-3}{jkr} + \cfrac{1}{\cfrac{2n-5}{jkr} + \ddots + \cfrac{1}{\cfrac{3}{jkr} + \cfrac{1}{\cfrac{1}{jkr} + 1}}}}}} \right\} . \qquad (5.81)$$

These continued fraction expansions describe the *Cauer canonic representations* of the outward propagating TM modes (Figure 5.16) and TE modes (Figure 5.17). We note that the equivalent circuit representing the TE_{mn} mode is dual to the the equivalent circuit representing the TM_{mn} mode. The equivalent circuits for the radiation modes exhibit high–pass character. For very low frequencies the wave impedance of the TM $_{mn}$ mode is represented by a capacitor $C_{0n} = \varepsilon r/n$ and the characteristic impedance of the TE_{mn} mode is represented by an inductor $L_{0n} = \mu r/n$. For $f \to \infty$ we obtain $Z_{mn}^{+TM}, Z_{mn}^{+TE} \to \eta$.

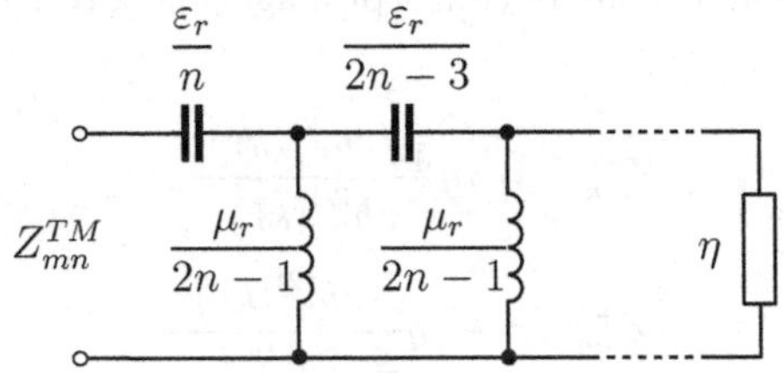

Fig. 5.16. Equivalent circuit of TM_{mn} spherical wave.

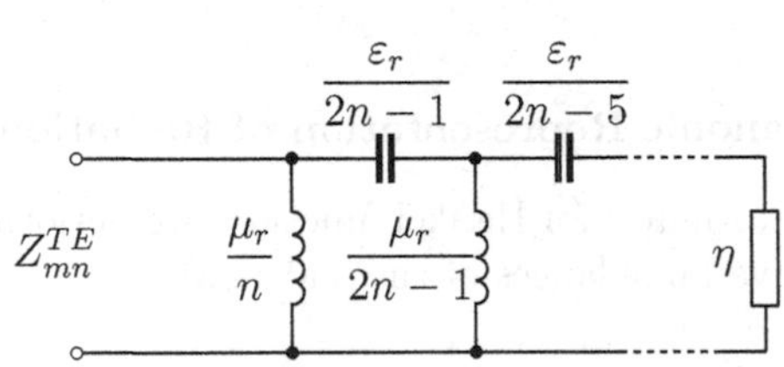

Fig. 5.17. Equivalent circuit of TE_{mn} spherical wave.

V.2 The Complete Equivalent Circuit of Radiating Electromagnetic Structures

In order to establish the equivalent circuit of a reciprocal linear lossless radiating electromagnetic structure, we embed the structure in a sphere $\mathcal{S}$ according to Figure 5.18.

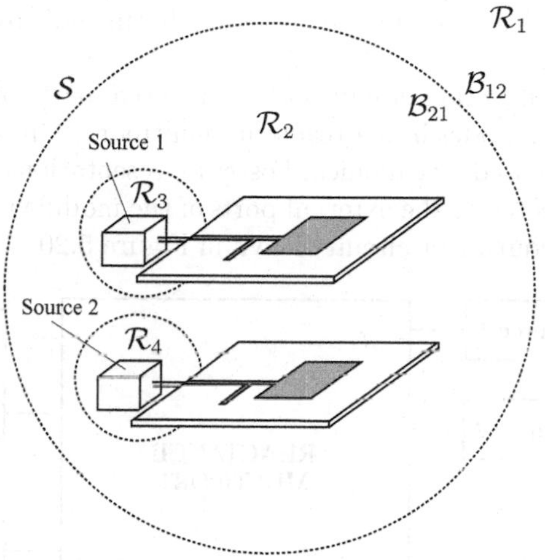

Fig. 5.18. The complete radiating electromagnetic structure.

The internal sources 1 and 2 are enclosed in regions $\mathcal{R}_3$ and $\mathcal{R}_4$. Region $\mathcal{R}_2$ only contains the reciprocal passive linear electromagnetic structure. Region $\mathcal{R}_1$ is the the infinite free space region outside the sphere $\mathcal{S}$. $\mathcal{R}_2$ may be either considered as a whole or may be subdivided into subregions. If $\mathcal{R}_2$ is considered as a whole it may be modeled either by a canonical Foster admittance representation according to Figure 5.12 or a canonical Foster impedance representation according to Figure 5.14. If the internal sources are coupled via a single transverse mode with the electromagnetic structure one port per source is required to model the coupling between the source and the electromagnetic structure. The radiating modes in $\mathcal{R}_1$ are represented by one–ports modeled by canonical Cauer representations according to Figure 5.16 and Figure 5.17 respectively. The external ports of the canonical Foster equivalent circuit, i.e. the ports representing the tangential field on the surface of $\mathcal{S}$ are connected via a connection network as shown in Figure 5.4.
From the above considerations we obtain for a reciprocal linear lossless radiating electromagnetic structure with internal sources an equivalent circuit described by a block diagram as shown in Figure 5.19. This block structure can be further simplified by contracting the equivalent circuit describing the electromagnetic

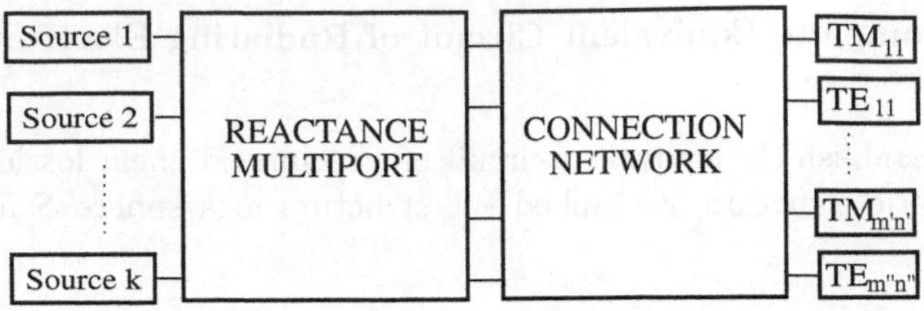

Fig. 5.19. Equivalent circuit of the complete radiating electromagnetic structure.

structure $\mathcal{R}_2$, the connection circuit and the reactive parts of the equivalent circuits of the radiation modes into a reactance multiport. This reactance multiport again may be represented by canonical Foster representations. Now the remaining resistors η are connected to the external ports of the modified reactance multiport and we obtain the equivalent circuit shown in Figure 5.20.

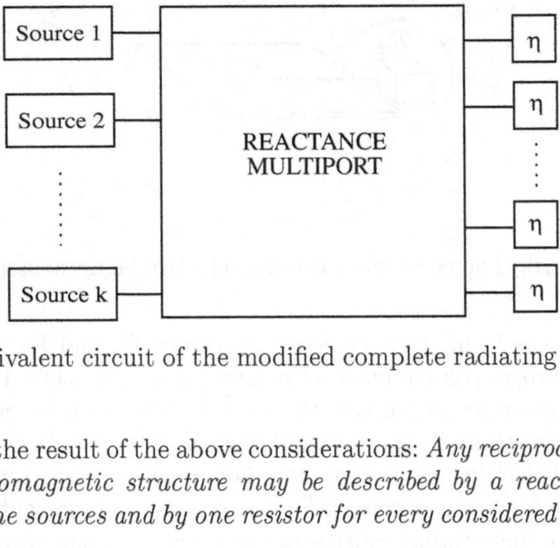

Fig. 5.20. Equivalent circuit of the modified complete radiating electromagnetic structure.

We summarize the result of the above considerations: *Any reciprocal linear lossless radiating electromagnetic structure may be described by a reactance multiport, terminated by the sources and by one resistor for every considered radiation mode.*

VI Solving the Entire Field Problem via Tableau Equations

VI.1 Primary and Secondary Fields

With reference to Figure 5.21, consider a subdomain $\mathcal{R}$, enclosed by a boundary $\mathcal{B}$ which consists partially of p.e.c. (surfaces $\mathcal{S}_m$) and partially of ports (openings) extending over surfaces $\mathcal{S}_o$. Such subdomains can also extend to infinity and, in this case, two different possibilities arise, depending on whether the surface at infinity is considered as a port or as a surface $\mathcal{S}_\infty$ where a boundary condition is imposed. In the latter case, we assume a Sommerfeld radiation condition for the fields on this surface. We also assume that, if sources are present inside the subdomain $\mathcal{R}$, they are contained inside a volume $\mathcal{V}'$, which is enclosed by a portion of the surface $\mathcal{S}_o$; the remaining volume $\mathcal{V}$, bounded by the surfaces $\mathcal{S}_m$, $\mathcal{S}_o$ and, possibly, $\mathcal{S}_\infty$ is therefore a source-free region.

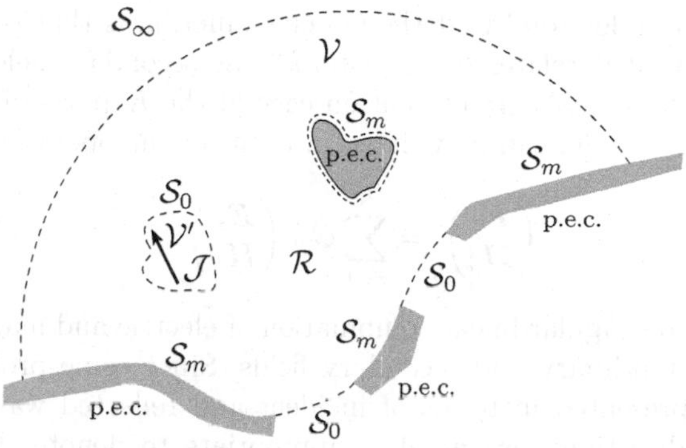

Fig. 5.21. A subdomain with openings described by surfaces $\mathcal{S}_o$ on which primary and secondary fields are specified; each opening is a port denoted by an index k. The surfaces $\mathcal{S}_m$ conform with p.e.c. boundaries and the Sommerfeld radiation condition is imposed on the surface $\mathcal{S}_\infty$ at infinity.

In order to parameterize the subdomain electromagnetically we introduce on the ports $\mathcal{S}_o$ *primary fields* (input quantities) and *secondary fields* (output quantities), denoted by subscripts p and s, respectively. In particular, we denote the tangential components of primary and secondary fields on a particular port k of subdomain $\mathcal{R}$ by the vectors

$$\begin{pmatrix} \boldsymbol{E}_k(\boldsymbol{\rho}) \\ \boldsymbol{H}_k(\boldsymbol{\rho}) \end{pmatrix}_p , \qquad \begin{pmatrix} \boldsymbol{E}_k(\boldsymbol{\rho}) \\ \boldsymbol{H}_k(\boldsymbol{\rho}) \end{pmatrix}_s . \tag{5.82}$$

The dependence on the two-dimensional vector $\boldsymbol{\rho}$ tangential to the boundary surface will be omitted subsequently in the notation unless clarity requires it. Primary fields are imposed on boundaries $\mathcal{S}_o$ and generate secondary fields inside the subdomain under consideration, the total field in the subdomain being expressed as the superposition of primary and secondary fields. Secondary fields may be expressed in terms of primary fields by introducing an operator $\hat{G}_{jk}$ which provides the secondary field on port j when *only the primary field on port k* is present. In fact, the secondary fields at port j in region $\mathcal{R}$ can be written in terms of the primary fields at the other ports by making use of (2.241) and (2.242)

$$\begin{pmatrix} \boldsymbol{E}_j(\boldsymbol{\rho}) \\ \boldsymbol{H}_j(\boldsymbol{\rho}) \end{pmatrix}_s = \sum_{k=1}^{K} \int_{\mathcal{B}_{\ell k}} d\boldsymbol{\rho}' \begin{pmatrix} \nabla \times \hat{G}_{jk}(\boldsymbol{\rho},\boldsymbol{\rho}') & j\omega\mu \hat{G}_{jk}(\boldsymbol{\rho},\boldsymbol{\rho}') \\ -j\omega\varepsilon \hat{G}_{jk}(\boldsymbol{\rho},\boldsymbol{\rho}') & \nabla \times \hat{G}_{jk}(\boldsymbol{\rho},\boldsymbol{\rho}') \end{pmatrix} \cdot \begin{pmatrix} \boldsymbol{E}_k(\boldsymbol{\rho}') \\ \boldsymbol{H}_k(\boldsymbol{\rho}') \end{pmatrix}_p . \tag{5.83}$$

In (5.83), $\hat{G}_{jk}(\boldsymbol{\rho},\boldsymbol{\rho}')$ is the appropriate dyadic Green's function for subdomain $\mathcal{R}$. It is noted that the electric and magnetic fields on the left side of (5.83) are those on port j, while the electric and magnetic fields on right side are those on port k. Similarly, it is understood that the Green's function is that pertaining to the subdomain and that it relates ports j and k. The secondary fields generated on port j when primary fields are present on each of the K ports of the subdomain $\mathcal{R}$, hence for $k = 1 \dots K$, can be written in terms of the operator $\hat{G}_{jk}$ as

$$\begin{pmatrix} \boldsymbol{E}_j \\ \boldsymbol{H}_j \end{pmatrix}_s = \sum_{k=1}^{K} \hat{G}_{jk} \begin{pmatrix} \boldsymbol{E}_k \\ \boldsymbol{H}_k \end{pmatrix}_p . \tag{5.84}$$

In general, any nonsingular linear combination of electric and magnetic fields can be considered for primary and secondary fields. Since some problems are more satisfactorily represented in terms of incident and reflected waves, or in terms of cascaded unidirectional waves, it is appropriate to denote the primary and secondary fields by the vectors $(\boldsymbol{F}_k)_{p,s}$ which are obtained from the primary and secondary fields in (5.84) by the transformation

$$\begin{pmatrix} (\boldsymbol{F}_k)_p \\ (\boldsymbol{F}_k)_s \end{pmatrix} = \begin{pmatrix} \xi_{11} & \xi_{12} \\ \xi_{21} & \xi_{22} \end{pmatrix} \begin{pmatrix} \boldsymbol{E}_k \\ \mathbf{H}_k \end{pmatrix} \tag{5.85}$$

with the matrix elements ξ_{ij} matched to the desired representation. Finally, we introduce the transformation operators $\left(\hat{T}_k\right)_p$ which allow (5.85) to be written as

$$\begin{pmatrix} (\boldsymbol{F}_k)_p \\ (\boldsymbol{F}_k)_s \end{pmatrix} = \left(\hat{T}_k\right)_{p,s} \begin{pmatrix} \boldsymbol{E}_k \\ \boldsymbol{H}_k \end{pmatrix}_{p,s} . \tag{5.86}$$

Then combining (5.86) and (5.84), we may write the relations between primary and secondary field state vectors in compact operator form as

$$(\boldsymbol{F}_j)_s = \sum_{k=1}^{K} \hat{O}_{jk} \, (\boldsymbol{F}_k)_p \tag{5.87}$$

with the operator $\hat{O}_{jk}$ obtained in terms of the operator $\hat{G}_{jk}$ as

$$\hat{O}_{jk} = \left(\hat{T}_j\right)_s \hat{G}_{jk} \left(\hat{T}_k\right)_p^{-1} \tag{5.88}$$

and the superscript (-1) denoting the inverse. These "*domain*" field relations are the field analog of "domain" relations in network theory, where circuit elements are described by their constitutive relations (e.g. resistors, capacitors, etc.) sometimes also referred to as branch equations.

VI.2 Choice of Primary and Secondary Fields for a Subregion

From the previous discussion it is apparent that the tangential components of the electric and magnetic fields on the apertures provide the quantities of interest for an input-output representation.
For that subregion the state is represented by the field inside the subdomain. For the moment we postpone the discussion on how to represent the state inside a subregion. We note, however, that the state is generated by the selected input variables and, in turn, provide the output quantities.
The selection of the input-output variables can be effected in a variety of ways. In general, for the subdomain with K ports the relevant field quantities are the transverse component of the electric and magnetic fields on the apertures:

$$\left[\boldsymbol{E}_1, \boldsymbol{E}_2, \ldots, \boldsymbol{E}_k, \ldots, \boldsymbol{E}_K, \boldsymbol{H}_1, \boldsymbol{H}_2, \ldots, \boldsymbol{H}_k, \ldots, \boldsymbol{H}_K\right]^T . \tag{5.89}$$

Half of these quantities may be taken as primary (input) fields, while the other half forms the secondary (output) field variables. As an example, by selecting the electric fields as input quantities we obtain an admittance representation (transfer function) in terms of standing waves of the form:

$$\left[\boldsymbol{H}_1, \boldsymbol{H}_2, \ldots, \boldsymbol{H}_k, \ldots, \boldsymbol{H}_K\right]^T = \hat{Y} \cdot \left[\boldsymbol{E}_1, \boldsymbol{E}_2, \ldots, \boldsymbol{E}_k, \ldots, \boldsymbol{E}_K\right]^T . \tag{5.90}$$

Alternatively we may select the magnetic field variables as primary fields, hence obtaining a standing wave representation of the impedance type:

$$\left[\boldsymbol{E}_1, \boldsymbol{E}_2, \ldots, \boldsymbol{E}_k, \ldots, \boldsymbol{E}_K\right]^T = \hat{Z} \cdot \left[\boldsymbol{H}_1, \boldsymbol{H}_2, \ldots, \boldsymbol{H}_k, \ldots, \boldsymbol{H}_K\right]^T . \tag{5.91}$$

Clearly, several other representations are possible, including the scattering representation in terms of incident and reflected waves and the transfer (ABCD) representations in terms of unidirectional waves.
Finally, it is noted that the above formalism accommodates description of the subdomain electrical *network* behavior by direct analogy with its field behavior. Thus, by using the analogy between tangential electric fields and voltages, and between tangential magnetic fields and currents, respectively, it is possible to obtain for each subdomain a network description with network elements represented by operators.

VI.3 A Constraint on the Choice of Primary and Secondary Fields

Concerning the choice of primary and secondary fields for the connection network, it is immediately apparent from (5.21a) and (5.21b) that we cannot choose the electric fields, $\mathbf{E}_t^\alpha$ and $\mathbf{E}_t^\beta$, on both sides as primary fields for the connection network. This choice is not self-consistent, and it does not relate the primary fields $\mathbf{E}_t^\alpha$ and $\mathbf{E}_t^\beta$ to the secondary fields $\mathbf{H}_t^\alpha$ and $\mathbf{H}_t^\beta$. In a similar manner we cannot select the magnetic field $\mathbf{H}_t^\alpha$ and $\mathbf{H}_t^\beta$ at both sides as primary fields and the electric fields $\mathbf{E}_t^\alpha$ and $\mathbf{E}_t^\beta$ as secondary fields. In other words *the connection network cannot be represented by an admittance or an impedance network*, since it corresponds to a volume of zero measure and does not contain any element which can either store or dissipate energy, as shown by the field form of Tellegen's theorem.

Other constraints on the choice of primary and secondary fields will appear after we introduce a finite expansion set for discretizing the connection relationships, as illustrated in the next section.

VI.4 Topological Relationships: Operator Form

An interface between two adjacent regions $\mathcal{R}_\ell$ and $\mathcal{R}_k$ has two different boundaries, which we have denoted by $\mathcal{B}_{\ell k}$ and $\mathcal{B}_{k\ell}$, one pertaining to region $\mathcal{R}_\ell$ and the other pertaining to region $\mathcal{R}_k$, respectively, as illustrated in Figure 5.2. Generically, we shall identify with superscripts α the quantities pertaining to the boundaries $\mathcal{B}_{\ell k}$ separating two regions for which $k > \ell$, and with superscripts β the quantities pertaining to the boundaries $\mathcal{B}_{k\ell}$ separating two regions for which $\ell > k$. Accordingly, we reorder the primary and secondary fields as

$$(\mathbf{F})_p = \begin{bmatrix} (\mathbf{F})_p^\alpha \\ (\mathbf{F})_p^\beta \end{bmatrix}, \quad (\mathbf{F})_s = \begin{bmatrix} (\mathbf{F})_s^\alpha \\ (\mathbf{F})_s^\beta \end{bmatrix} . \tag{5.92}$$

Adopting this partitioning we have

$$\begin{bmatrix} (\mathbf{F})_s^\alpha \\ (\mathbf{F})_s^\beta \end{bmatrix} = \begin{bmatrix} \hat{O}^{\alpha\alpha} & \hat{O}^{\alpha\beta} \\ \hat{O}^{\beta\alpha} & \hat{O}^{\beta\beta} \end{bmatrix} \begin{bmatrix} (\mathbf{F})_p^\alpha \\ (\mathbf{F})_p^\beta \end{bmatrix} . \tag{5.93}$$

The transverse electric and magnetic fields $\mathbf{E}_t$ and $\mathbf{H}_t$ at all the boundaries connecting different regions are unique and may be obtained from the primary and secondary state vectors. Let us introduce the operators $\hat{C}^{\alpha,\beta}$ which transform the state vectors into the transverse electromagnetic fields at that boundary. The following transformations hold on the boundaries α, β

$$\hat{C}^{\alpha,\beta} \begin{bmatrix} (\mathbf{F})_p^{\alpha,\beta} \\ (\mathbf{F})_s^{\alpha,\beta} \end{bmatrix} = \begin{bmatrix} \mathbf{E}_t \\ \mathbf{H}_t \end{bmatrix} . \tag{5.94}$$

Invoking continuity of the transverse electric and magnetic fields yields the following relations

$$\hat{C}^\alpha \begin{bmatrix} (\mathbf{F})_p^\alpha \\ (\mathbf{F})_s^\alpha \end{bmatrix} - \hat{C}^\beta \begin{bmatrix} (\mathbf{F})_p^\beta \\ (\mathbf{F})_s^\beta \end{bmatrix} = 0 . \tag{5.95}$$

The above equations represent the *connection relations* describing the connection between the different regions into which the entire problem space has been partitioned. Note that (5.95) represents the continuity of the tangential components of the electric and magnetic fields and is therefore the field theoretic analog of the Kirchhoff voltage and current laws. The *subdomain relations* in (5.93) and the *connection relations* in (5.95) together constitute the formal solution of the entire field problem.

VI.5 The Tableau Equations for Fields: Operator Form

Similar to what is done in circuit theory [15, p. 715], (5.93) and (5.95) can be assembled in the most general way by using the *Tableau representation* for the field problem. In particular, by employing the following form for the operators $\hat{C}^{\alpha,\beta}$

$$\hat{C}^{\alpha,\beta} = \begin{bmatrix} \hat{C}_{11}^{\alpha,\beta} & \hat{C}_{12}^{\alpha,\beta} \\ \hat{C}_{21}^{\alpha,\beta} & \hat{C}_{22}^{\alpha,\beta} \end{bmatrix} \tag{5.96}$$

we may rewrite the entire electromagnetic problem as

$$\begin{bmatrix} \hat{O}^{\alpha\alpha} & \hat{O}^{\alpha\beta} & -\tilde{\mathbf{I}} & \tilde{\mathbf{o}} \\ \hat{O}^{\beta\alpha} & \hat{O}^{\beta\beta} & \tilde{\mathbf{o}} & -\tilde{\mathbf{I}} \\ \hat{C}_{11}^{\alpha} & -\hat{C}_{11}^{\beta} & \hat{C}_{12}^{\alpha} & -\hat{C}_{12}^{\beta} \\ \hat{C}_{21}^{\alpha} & -\hat{C}_{21}^{\beta} & \hat{C}_{22}^{\alpha} & -\hat{C}_{22}^{\beta} \end{bmatrix} \cdot \begin{bmatrix} (\mathbf{F})_p^{\alpha} \\ (\mathbf{F})_p^{\beta} \\ (\mathbf{F})_s^{\alpha} \\ (\mathbf{F})_s^{\beta} \end{bmatrix} = 0\,. \tag{5.97}$$

In (5.97) the first two rows pertain to the subdomain equations while the last two rows describe the topological relations. In the terminology of circuit theory, the first two equations represent the branch equations, while the last two equations express Kirchhoff's laws.

VI.6 Solving the Entire Field Problem via Tableau Equations: Discretized Form

Numerical solution of the subdomain and connection equations requires discretization, for example, representation of the fields on bases which, for numerical implementation, constitute a finite set. It is thus appropriate to distinguish between the exact field representations of the previous sections denoted, for example, by $\mathbf{F}$, and the approximate fields considered in this section for numerical computations, denoted by $\tilde{\mathbf{F}}$.
To discretize the subdomain and connection equations we proceed in the customary fashion:

- Introduce the expansion function set
- Introduce the weight function set
- Apply the expansion and weight functions to the subdomain equations to obtain the *multiport networks* for the subdomain regions
- Apply the expansion and weight functions to the connection equations to obtain the *connection networks* between subdomains.

Although the actual choice of the expansion basis plays a fundamental role with respect to numerical convergence etc., we are concerned here only with its general properties, namely that the basis is complete and satisfies boundary, edge and radiation conditions when necessary.

VI.7 Field Discretization

Expansion Basis Function Set

With reference to (5.92) we expand the fields as

$$\left(\tilde{\mathbf{F}}\right)_p^\alpha = \sum_n^{N_p^\alpha} f_{pn}^\alpha \mathbf{d}_{pn}^\alpha(\boldsymbol{\rho}) \,, \tag{5.98a}$$

$$\left(\tilde{\mathbf{F}}\right)_p^\beta = \sum_n^{N_p^\beta} f_{pn}^\beta \mathbf{d}_{pn}^\beta(\boldsymbol{\rho}) \,, \tag{5.98b}$$

$$\left(\tilde{\mathbf{F}}\right)_s^\alpha = \sum_n^{N_s^\alpha} f_{sn}^\alpha \mathbf{d}_{sn}^\alpha(\boldsymbol{\rho}) \,, \tag{5.98c}$$

$$\left(\tilde{\mathbf{F}}\right)_s^\beta = \sum_n^{N_s^\beta} f_{sn}^\beta \mathbf{d}_{sn}^\beta(\boldsymbol{\rho}) \,. \tag{5.98d}$$

Here $\mathbf{d}_{pn}^\alpha(\boldsymbol{\rho})$ denotes the n-th basis function for the primary fields on the boundary of type α, and f_{pn}^α denotes the basis amplitude coefficient. Similar interpretations apply to quantities identified by superscripts β and/or subscripts s. In these equations, because the fields are truncated after a finite number of terms, the notation is that for approximate fields. Note that both $\mathbf{d}_{pn}^\alpha(\boldsymbol{\rho})$ and f_{pn}^α can, in principle, depend on other vector functions but this will not be pursued further because it is conceptually straightforward although notationally cumbersome.

Weighting Functions

In directly analogous fashion, we introduce the weighting functions

$$\begin{array}{ll} \mathbf{p}_{pm}^\alpha(\boldsymbol{\rho}) & m = 1, \ldots, N_p^\alpha \\ \mathbf{p}_{pm}^\beta(\boldsymbol{\rho}) & m = 1, \ldots, N_p^\beta \\ \mathbf{p}_{sm}^\alpha(\boldsymbol{\rho}) & m = 1, \ldots, N_s^\alpha \\ \mathbf{p}_{sm}^\beta(\boldsymbol{\rho}) & m = 1, \ldots, N_s^\beta \end{array} \tag{5.99}$$

which are used to test the subdomain and connection relationships.

Discretization of the Subdomain Equations

By inserting the expansions for the primary fields in (5.98b) into the subdomain relations (5.93) and testing the resulting equations with $\mathbf{p}^{\alpha}_{sm}(\boldsymbol{\rho})$ and $\mathbf{p}^{\beta}_{sm}(\boldsymbol{\rho})$, respectively, we obtain

$$\begin{bmatrix} \int_{\mathcal{B}_\alpha} dA\, \mathbf{p}^{\alpha}_{sm}(\boldsymbol{\rho})\, (\mathbf{F})^{\alpha}_{s}(\boldsymbol{\rho}) \\ \int_{\mathcal{B}_\beta} dA\, \mathbf{p}^{\beta}_{sm}(\boldsymbol{\rho})\, (\mathbf{F})^{\beta}_{s}(\boldsymbol{\rho}) \end{bmatrix} =$$

$$\begin{bmatrix} \int_{\mathcal{B}_\alpha} dA\, \mathbf{p}^{\alpha}_{sm}(\boldsymbol{\rho}) & \widetilde{\mathbf{o}} \\ \widetilde{\mathbf{o}} & \int_{\mathcal{B}_\beta} dA\, \mathbf{p}^{\beta}_{sm}(\boldsymbol{\rho}) \end{bmatrix} \begin{bmatrix} \hat{O}^{\alpha\alpha} & \hat{O}^{\alpha\beta} \\ \hat{O}^{\beta\alpha} & \hat{O}^{\beta\beta} \end{bmatrix} \begin{bmatrix} \sum_{n}^{N_{p\alpha}} f^{\alpha}_{pn} \mathbf{d}^{\alpha}_{pn}(\boldsymbol{\rho}') \\ \sum_{n}^{N_{p\beta}} f^{\beta}_{pn} \mathbf{d}^{\beta}_{pn}(\boldsymbol{\rho}') \end{bmatrix}. \tag{5.100}$$

Next, we introduce the definitions

$$t^{\alpha}_{sm} = \int_{\mathcal{B}_\alpha} dA\, \mathbf{p}^{\alpha}_{sm}(\boldsymbol{\rho}) \mathbf{F}^{\alpha}_{s}(\boldsymbol{\rho}), \tag{5.101a}$$

$$t^{\beta}_{sm} = \int_{\mathcal{B}_\beta} dA\, \mathbf{p}^{\beta}_{sm}(\boldsymbol{\rho}) \mathbf{F}^{\beta}_{s}(\boldsymbol{\rho}) \tag{5.101b}$$

and

$$o^{\gamma\eta}_{mn} = \int_{\mathcal{B}_\alpha} dA\, \mathbf{w}^{\gamma}_{sm}(\boldsymbol{\rho}) \hat{O}^{\gamma\eta} \mathbf{f}^{\eta}_{pn}(\boldsymbol{\rho}') \tag{5.102}$$

with superscripts $\gamma, \eta = \alpha, \beta$. Forming the vectors

$$\mathbf{t}^{\alpha}_{s} = \begin{bmatrix} t^{\alpha}_{s1} \\ t^{\alpha}_{s2} \\ \vdots \\ t^{\alpha}_{sN_{s\alpha}} \end{bmatrix}, \qquad \mathbf{t}^{\beta}_{s} = \begin{bmatrix} t^{\beta}_{s1} \\ t^{\beta}_{s2} \\ \vdots \\ t^{\beta}_{sN_{s\beta}} \end{bmatrix}, \tag{5.103}$$

$$\mathbf{f}^{\alpha}_{p} = \begin{bmatrix} f^{\alpha}_{p1} \\ f^{\alpha}_{p2} \\ \vdots \\ f^{\alpha}_{pN_{p\alpha}} \end{bmatrix}, \qquad \mathbf{f}^{\beta}_{p} = \begin{bmatrix} f^{\beta}_{p1} \\ f^{\beta}_{p2} \\ \vdots \\ f^{\beta}_{pN_{p\beta}} \end{bmatrix}, \tag{5.104}$$

$$\mathbf{f}^{\alpha}_{s} = \begin{bmatrix} f^{\alpha}_{s1} \\ f^{\alpha}_{s2} \\ \vdots \\ f^{\alpha}_{sN_{s\alpha}} \end{bmatrix}, \qquad \mathbf{f}^{\beta}_{s} = \begin{bmatrix} f^{\beta}_{s1} \\ f^{\beta}_{s2} \\ \vdots \\ f^{\beta}_{sN_{s\beta}} \end{bmatrix}, \tag{5.105}$$

and the matrices

$$\widetilde{\mathbf{o}}^{\gamma\eta} = \begin{bmatrix} o_{11}^{\gamma\eta} \; o_{12}^{\gamma\eta} \cdots & \\ o_{21}^{\gamma\eta} \; o_{22}^{\gamma\eta} \cdots & \\ \vdots \quad \ddots & \\ & o_{N_{s\gamma}N_{s\eta}}^{\gamma\eta} \end{bmatrix} \tag{5.106}$$

permits writing of the discretized subdomain equations compactly as

$$\begin{bmatrix} \mathbf{t}_s^{\alpha} \\ \mathbf{t}_s^{\beta} \end{bmatrix} = \begin{bmatrix} \widetilde{\mathbf{o}}^{\alpha\alpha} & \widetilde{\mathbf{o}}^{\alpha\beta} \\ \widetilde{\mathbf{o}}^{\beta\alpha} & \widetilde{\mathbf{o}}^{\beta\beta} \end{bmatrix} \begin{bmatrix} \mathbf{f}_p^{\alpha} \\ \mathbf{f}_p^{\beta} \end{bmatrix} . \tag{5.107}$$

Discretization of the Connection Equations

The connection equations (5.95) are tested by using the weight functions $\mathbf{p}_{pm}^{\alpha}(\boldsymbol{\rho}), \mathbf{p}_{pm}^{\beta}(\boldsymbol{\rho})$ and by expanding primary and secondary fields according to (5.98b) and (5.98d). As a consequence, (5.95) becomes

$$\begin{bmatrix} \int_{\mathcal{B}_\alpha} dA\, \mathbf{p}_{pm}^{\alpha}(\boldsymbol{\rho}) & 0 \\ 0 & \int_{\mathcal{B}_\beta} dA\, \mathbf{p}_{pm}^{\beta}(\boldsymbol{\rho}) \end{bmatrix} \cdot \begin{bmatrix} C_{11}^{\alpha}(\boldsymbol{\rho}) & -C_{11}^{\beta}(\boldsymbol{\rho}) & C_{12}^{\alpha}(\boldsymbol{\rho}) & -C_{12}^{\beta}(\boldsymbol{\rho}) \\ C_{21}^{\alpha}(\boldsymbol{\rho}) & -C_{21}^{\beta}(\boldsymbol{\rho}) & C_{22}^{\alpha}(\boldsymbol{\rho}) & -C_{22}^{\beta}(\boldsymbol{\rho}) \end{bmatrix} \cdot \begin{bmatrix} \sum_n^{N_{p\alpha}} f_{pn}^{\alpha} \mathbf{d}_{pn}^{\alpha}(\boldsymbol{\rho}) \\ \sum_n^{N_{p\beta}} f_{pn}^{\beta} \mathbf{d}_{pn}^{\beta}(\boldsymbol{\rho}) \\ \sum_n^{N_{s\alpha}} f_{sn}^{\alpha} \mathbf{d}_{sn}^{\alpha}(\boldsymbol{\rho}) \\ \sum_n^{N_{s\beta}} f_{sn}^{\beta} \mathbf{d}_{sn}^{\beta}(\boldsymbol{\rho}) \end{bmatrix} = 0 .$$

Defining the matrix elements

$$c_{ijmn}^{\gamma\eta} = \int_{\mathcal{B}_\gamma} dA\, \mathbf{p}_{pm}^{\gamma}(\boldsymbol{\rho}) C_{ij}^{\eta}(\boldsymbol{\rho}) \mathbf{d}_{pn}^{\eta}(\boldsymbol{\rho}) \tag{5.108}$$

and forming the matrices

$$\widetilde{\mathbf{c}}_{ij}^{\gamma\eta} = \begin{bmatrix} c_{ij11}^{\gamma\eta} \; c_{ij12}^{\gamma\eta} \cdots & \\ c_{ij21}^{\gamma\eta} \; c_{ij22}^{\gamma\eta} \cdots & \\ \vdots \quad \ddots & \\ & c_{ijNN}^{\gamma\eta} \end{bmatrix} \tag{5.109}$$

leads to the discretized connection equations in (5.111). In (5.108) the indices $i, j = 1, 2$, and the first and second superscripts $\gamma, \eta = \alpha, \beta$ refer, respectively, to the weight function, and to the superscript of the expansion function.

VI.8 The Tableau Equations for Discretized Fields

Using lower case symbols to describe the discretized quantities, the subdomain equation (5.93) is written in its discretized form as (for details see Section VI.7)

$$\begin{bmatrix} \mathbf{f}_s^\alpha \\ \mathbf{f}_s^\beta \end{bmatrix} = \begin{bmatrix} \widetilde{\mathbf{o}}^{\alpha\alpha} & \widetilde{\mathbf{o}}^{\alpha\beta} \\ \widetilde{\mathbf{o}}^{\beta\alpha} & \widetilde{\mathbf{o}}^{\beta\beta} \end{bmatrix} \begin{bmatrix} \mathbf{f}_p^\alpha \\ \mathbf{f}_p^\beta \end{bmatrix} . \tag{5.110}$$

The matrix $\widetilde{\mathbf{o}}$ is the multiport equivalent network which is used in the actual numerical solution of the field problem. This network may be obtained in some cases directly by projection (overlap integrals) from the Green's functions of the different regions. It is therefore crucial to select the appropriate Green's function representations, both for preserving phenomenological insight and for numerical efficiency.

The connection equations (5.95) have been similarly discretized leading to the following set of equations

$$\begin{bmatrix} \widetilde{\mathbf{c}}_{11}^{\alpha\alpha} & \widetilde{\mathbf{c}}_{11}^{\alpha\beta} & \widetilde{\mathbf{c}}_{12}^{\alpha\alpha} & \widetilde{\mathbf{c}}_{12}^{\alpha\beta} \\ \widetilde{\mathbf{c}}_{21}^{\beta\alpha} & \widetilde{\mathbf{c}}_{21}^{\beta\beta} & \widetilde{\mathbf{c}}_{22}^{\beta\alpha} & \widetilde{\mathbf{c}}_{22}^{\beta\beta} \end{bmatrix} \begin{bmatrix} \mathbf{f}_p^\alpha \\ \mathbf{f}_p^\beta \\ \mathbf{f}_s^\alpha \\ \mathbf{f}_s^\beta \end{bmatrix} = 0 \tag{5.111}$$

which defines the discretized *connection network*. From (5.111) we may express the secondary fields in terms of the primary fields and, upon substitution into (5.107), achieve solution of the field problem. This procedure may be convenient in some particular cases, but the most general approach is achieved by using the Tableau analysis.

Combining (5.110) and (5.111), we obtain the discretized form of the Tableau equations,

$$\begin{bmatrix} \widetilde{\mathbf{o}}^{\alpha\alpha} & \widetilde{\mathbf{o}}^{\alpha\beta} & -\widetilde{\mathbf{1}} & \widetilde{\mathbf{o}} \\ \widetilde{\mathbf{o}}^{\beta\alpha} & \widetilde{\mathbf{o}}^{\beta\beta} & \widetilde{\mathbf{o}} & -\widetilde{\mathbf{1}} \\ \widetilde{\mathbf{c}}_{11}^{\alpha\alpha} & \widetilde{\mathbf{c}}_{11}^{\alpha\beta} & \widetilde{\mathbf{c}}_{12}^{\alpha\alpha} & \widetilde{\mathbf{c}}_{12}^{\alpha\beta} \\ \widetilde{\mathbf{c}}_{21}^{\beta\alpha} & \widetilde{\mathbf{c}}_{21}^{\beta\beta} & \widetilde{\mathbf{c}}_{22}^{\beta\alpha} & \widetilde{\mathbf{c}}_{22}^{\beta\beta} \end{bmatrix} \begin{bmatrix} \mathbf{f}_p^\alpha \\ \mathbf{f}_p^\beta \\ \mathbf{f}_s^\alpha \\ \mathbf{f}_s^\beta \end{bmatrix} = 0 . \tag{5.112}$$

As noted in [15, p. 225] we have as many Tableau equations as there are variables; thus, the price paid for this completely general approach is that the *Tableau analysis* involves many more equations than other possible but specific formulations. However, in the solution of complex structures, this fact turns out to be a blessing in disguise because the associated matrix is generally extremely sparse, thereby allowing use of highly efficient numerical algorithms.

References

[1] R. Mittra, T. Itoh, and T. S. Li, "Analytical and numerical studies of the relative convergence phenomenon arising in the solution of an integral equation by the moment method," *IEEE Trans. Microwave Theory Tech.*, vol. 20, no. 7, pp. 96–104, Jul. 1972.

[2] G. V. Eleftheriades, A. S. Omar, L. P. Katehi, and G. M. Rebeiz, "Some important properties of waveguide junction generalized scattering matrices in the context of the mode-matching technique," *IEEE Trans. Microwave Theory Tech.*, vol. 42, no. 10, pp. 1896–1903, Oct. 1994.

[3] R. Schmidt and P. Russer, "Modeling of cascaded coplanar waveguide discontinuities by the mode–matching approach," *IEEE Trans. Microwave Theory Tech.*, vol. 43, no. 12, pp. 2910–2917, Dec. 1995.

[4] L. B. Felsen, M. Mongiardo, and P. Russer, "Electromagnetic field representations and computations in complex structures I: Complexity architecture and generalized network formulation," *Int. J. Numerical Modelling: El. Networks, Devices and Fields*, vol. 15, pp. 93–107, 2002.

[5] P. Russer, M. Mongiardo, and L. B. Felsen, "Electromagnetic field representations and computations in complex structures III: Network representations of the connection and subdomain circuits," *Int. J. Numerical Modelling: El. Networks, Devices and Fields*, vol. 15, pp. 127–145, 2002.

[6] M. Righi, W. J. R. Hoefer, M. Mongiardo, and R. Sorrentino, "Efficient TLM diakoptics for separable structures," *IEEE Trans. Microwave Theory Tech.*, vol. 43, no. 4, pp. 854–859, Apr. 1995.

[7] M. Mongiardo and R. Sorrentino, "Efficient and versatile analysis of microwave structures by combined mode matching and finite difference methods," *IEEE Microwave Guided Wave Lett.*, vol. 3, no. 7, pp. 241–243, Aug. 1993.

[8] R. Harrington, *Field Computation by Moment Methods*, 2nd ed. Florida: Robert E. Krieger Publishing Company, 1982.

[9] R. Mittra and S.W. Lee, *Analytical Techniques in the Theory of Guided Waves*, 1st ed., ser. MacMillan Series in Electrical Science. New York: The MacMillan Company, 1971.

[10] J. Wang, *Generalized Moment Methods in Electromagnetics*, 1st ed. New York: John Wiley & Sons, Inc., 1991.

[11] R. F. Harrington, *Field Computation by Moment Methods.* New York: IEEE Press, 1993.

[12] A. Peterson, S.L. Ray and R. Mittra, *Computational Methods for Electromagnetics*, 1st ed., ser. IEEE/OUP Series on Electromagnetic Waves. New York: IEEE/OUP Press, 1998.

[13] A. Quarteroni, R. Sacco and F. Saleri, *Numerical Mathematics*, 1st ed., ser. Texts in Applied Mathematics. New York: Springer-Verlag, 2000.

[14] D. Dudley, *Mathematical Foundations for Electromagnetic Theory*, 1st ed., ser. IEEE-/OUP Series on Electromagnetic Wave Theory. New York: IEEE/OUP Press, 1994.

[15] L. Chua, C. Desoer, and E. Kuh, *Linear and Nonlinear Circuits.* New York: Mc Graw Hill, 1987.

[16] S. Hassani, *Mathematical Physics.* Berlin: Springer, 2002.

[17] V. Belevitch, *Classical network theory.* San Francisco, California: Holden-Day, 1968.

[18] R. F. Harrington, *Time Harmonic Electromagnetic Fields.* New York: McGraw-Hill, 1961.

[19] R. E. Collin, *Field Theory of Guided Waves.* New York: IEEE Press, 1991.

[20] W. Cauer, *Theorie der linearen Wechselstromschaltungen.* Berlin: Akademie-Verlag, 1954.

[21] L. Chu, "Physical limitations of omni–directional antennas," *J. Appl. Physics*, pp. 1163–1175, Dec. 1948.

[22] L. Felsen and N. Marcuvitz, *Radiation and Scattering of Waves.* Englewood Cliffs, NJ: Prentice Hall, 1972.

Appendix

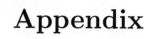

List of Symbols

Symbol	Description	Reference
a	waveguide width	(4.3)
A	field component transverse to the radial distance r	(2.206)
$\boldsymbol{A}$	magnetic vector potential	(2.74)
A	array	(5.9)
$\widetilde{\mathsf{A}}$	matrix	(5.31)
$\widetilde{\mathsf{A}}_\lambda$	real frequency-independent rank 1 matrix	(5.67)
A_α	coefficients with index α	(2.158)
α	index	(2.103)
b	physical dimension	(4.25)
$\boldsymbol{B}(\boldsymbol{r},t)$	magnetic flux density	(2.1a)
$\widetilde{\mathsf{B}}$	matrix	(5.31)
$\widetilde{\mathsf{B}}_\lambda$	real frequency-independent rank 1 matrix	(5.71)
$B_n(x)$	B-spline (bell-spline)	(5.14)
β	index	(4.25)
C	contour	(2.5)
$\bar{c}$	constant	(2.195)
$\bar{C}$	constant	(2.180)
$\hat{C}^{\alpha,\beta}$	operator	(5.94)
$\widetilde{\mathsf{C}}$	matrix	(5.31)
$\widetilde{\mathsf{c}}_{ij}^{\gamma\eta}$	matrices	(5.108)
c	propagation speed in a transmission line	(5.46a)
c_0	free–space propagation speed	(2.92)
C_p	capacitance	(5.52a)
C_{sn}	capacitance	(5.53b)
$\mathbf{d}_{pn}^{\alpha}(\boldsymbol{\rho})$	n-th basis function	(5.98d)
$\boldsymbol{D}(\boldsymbol{r},t)$	electric flux density	(2.1b)
$\widetilde{\mathsf{D}}$	matrix	(5.31)
$D_\varepsilon^2(z)$	derivative operator	(3.93a)
$D_\mu^2(z)$	derivative operator	(3.93b)
$\delta_{\alpha\beta}$	Kronecker delta	(2.145)
$\delta(u-u')$	delta function	(2.153)

Symbol	Description	Reference
$\boldsymbol{E}(\boldsymbol{r},t)$	electric field strength	(2.1a)
$\boldsymbol{E}(\boldsymbol{r})$	complex electric field strength in the frequency domain	(2.11)
$\delta\boldsymbol{E}$	difference of two electric fields	(2.59)
$\boldsymbol{E}_t^\alpha$	transverse electric field on side α	(5.21a)
$\boldsymbol{E}_t^\beta$	transverse electric field on side β	(5.21a)
$\boldsymbol{e}_n^\alpha$	*n-th* basis functions for electric field side on side α	(5.22b)
$\boldsymbol{e}_n^\beta$	*n-th* basis functions for electric field side on side β	(5.22b)
$\boldsymbol{e}_i{}'(\boldsymbol{\rho})$	TM orthonormal transverse vector eigenfunctions	(3.29)
$\boldsymbol{e}_i{}''(\boldsymbol{\rho})$	TE orthonormal transverse vector eigenfunctions	(3.37a)
ϵ	infinitesimal increment	(2.184)
ε	electric permittivity	(2.18b)
ε_i	imaginary part of electric permittivity	(2.61)
ε_0	free–space permittivity	(2.19b)
ε_r	real part of electric permittivity	(2.61)
$\underline{\underline{\varepsilon}}$	permittivity tensor	(2.21b)
η	free space impedance	(2.69)
$\overleftarrow{f}$	SL solution satisfying boundary condition on the left side	(2.175)
$\overrightarrow{f}$	SL solution satisfying boundary condition on the right side	(2.175)
$F(u)$	a function of (u)	(2.130)
$\bar{F}(u)$	functions replacing $f_\alpha(u)$	(2.130)
$\boldsymbol{F}$	electric vector potential	(2.84b)
$(\boldsymbol{F}_k)_{p,s}$	primary and secondary fields	(5.85)
$f_\alpha(u)$	eigenfunction	(2.127)
F_β	*β-th* expansion function	(5.4)
f_{pn}^α	basis amplitude coefficient	(5.98d)
$\mathbf{f}_p^\xi$	$\xi = \alpha, \beta$ array	(5.104)
$\mathbf{t}_s^\xi$	$\xi = \alpha, \beta$ array	(5.105)
$\gamma_{T1,2}$	constants	(2.94)
$\overrightarrow{\Gamma}$	parameter	(2.177)
$\overleftarrow{\Gamma}$	parameter	(2.177)
$g(u,u';\lambda)$	one-dimensional Green's function	(2.165)
G'	scalar TM Green's function	(3.84a)
G''	scalar TE Green's function	(3.85)
$\mathcal{G}_e$	dyadic Green's function (electric type)	(2.236)
$\mathcal{G}_m$	dyadic Green's function (magnetic type)	(2.242)
$\hat{G}_{jk}$	operator relating ports jk	(5.83)

Symbol	Description	Reference
$\boldsymbol{H}(\boldsymbol{r},t)$	magnetic field strength	(2.1b)
$\delta\boldsymbol{H}$	difference of two magnetic fields	(2.59)
$\boldsymbol{H}_t^\alpha$	transverse magnetic field on side α	(5.21b)
$\boldsymbol{H}_t^\beta$	transverse magnetic field on side β	(5.21b)
$\boldsymbol{h}_n^\alpha$	*n-th* basis functions for magnetic field side on side α	(5.22d)
$\boldsymbol{h}_n^\beta$	*n-th* basis functions for magnetic field side on side β	(5.22d)
$\boldsymbol{h}_i{}'(\boldsymbol{\rho})$	TM orthonormal transverse vector eigenfunctions	(3.29)
$\boldsymbol{h}_i{}''(\boldsymbol{\rho})$	TE orthonormal transverse vector eigenfunctions	(3.37a)
$H_\nu^{(1)}(k_0r)$	cylindrical Hankel function of first type	(2.125)
$H_\nu^{(2)}(k_0r)$	cylindrical Hankel function of second type	(2.125)
$h_\nu^{(1)}(k_0r)$	spherical Hankel function of first type	(2.117)
$h_\nu^{(2)}(k_0r)$	spherical Hankel function of second type	(2.117)
$I(t)$	current	(2.7)
I_n^ξ	$\xi=\alpha,\beta$, field amplitudes of the magnetic fields	(5.22d)
I	array for the expansions coefficients of the magnetic field	(5.25)
I^ξ	$\xi=\alpha,\beta$, arrays for the expansions coefficients of the magnetic field	(5.24)
$I_i'(z)$	TM modal currents	(3.31)
$I_i''(z)$	TE modal currents	(3.40)
$i_i'(z)$	TM modal currents	(3.43c)
$i_i''(z)$	TE modal currents	(3.43c)
$\mathscr{I}$	identity dyadic	(2.236)
$\widetilde{\mathbf{1}}$	identity matrix	(5.43b)
$J_\nu(k_0r)$	cylindrical Bessel function	(2.125)
$j_\nu(k_0r)$	spherical Bessel function	(2.117)
$\boldsymbol{J}(\boldsymbol{r},t)$	electric current density	(2.1b)
$\boldsymbol{J}_0$	impressed current density	(2.33)
$\boldsymbol{J}_s$	equivalent electric currents	(2.67b)
$\boldsymbol{J}^0$	constant vector	(3.69)
k	wavenumber	(3.3)
$\widetilde{\mathsf{K}}$	matrix	(5.34b)
k_0	free–space wavenumber	(2.91)
k_t	transverse wavenumber	(3.25)
$\bar{k}(z)$	modified wavenumber	(3.97)
κ	longitudinal wavenumber	(3.25)

Symbol	Description	Reference
ℓ	line element vector	(2.5)
l	transmission line length	Figure 5.7
λ	spectral parameter	(2.128)
λ_α	eigenvalue	(2.127)
$\widetilde{\mathbf{\Lambda}}$	matrix with elements Λ^{ξ}_{mn}	(5.28)
Λ^{ξ}_{mn}	$\xi = \alpha, \beta$, matrix element mn	(5.27)
$L(u)$	Sturm-Liouville operator	(2.128)
$\hat{L}(u)$	linear operator	(5.1)
$\widetilde{\mathbf{L}}$	matrix	(5.10)
L_1	vector operators	(3.92e)
L_2	vector operators	(3.92e)
L'_1	vector operators	(3.91a)
L'_2	vector operators	(3.91a)
$L_{\alpha\beta}$	matrix elements	(5.6)
L_0	inductances	(5.52b)
LL_s	inductances	(5.52b)
L_{pn}	inductances	(5.53a)
$\boldsymbol{M}(\boldsymbol{r})$	magnetic current density	(2.14a)
$\boldsymbol{M}^0$	constant vector	(3.69)
$\boldsymbol{M}_s$	equivalent magnetic currents	(2.67b)
$\widetilde{\mathbf{M}}$	matrix	(5.42)
μ	magnetic permeability	(2.18b)
μ_i	imaginary part of magnetic permeability	(2.61)
μ_0	free–space permeability	(2.19b)
μ_r	real part of magnetic permeability	(2.61)
$\underline{\underline{\mu}}$	permeability tensor	(2.21b)
$\mathbf{n}$	the outward unit vector normal to S	(2.4)
$N_\nu(k_0 r)$	cylindrical Neumann function	(2.125)
$n_\nu(k_0 r)$	spherical Neumann function	(2.117)
ν	order of Bessel function	(2.117)
$\boldsymbol{\nu}(s)$	outward normal unit vector	Figure (3.1)
$\widetilde{\mathbf{o}}$	zero matrix	(5.97)
$\tilde{O}_{jk}$	operator	(5.87)
$\widetilde{\mathbf{o}}^{\gamma\eta}$	matrices	(5.106)
ω	angular frequency	(2.11)
ω_1	angular frequency in a transmission line	(5.47)
ω_{0pn}	angular frequency	(5.54a)
ω_{0sn}	angular frequency	(5.54b)
ω_λ	angular frequency	(5.58)

Symbol	Description	Reference
p	positive real function	(2.127)
$\mathbf{p}_{pm}^{\alpha}(\boldsymbol{\rho})$	weighting functions	(5.99)
$P_{\nu}^{-\mu}(\cos\theta)$	associated Legendre functions	(2.116)
$\boldsymbol{\Pi}_e$	electric Hertz vector potential	(2.87b)
Π_e	scalar electric potential	(3.19)
$\boldsymbol{\Pi}_h$	magnetic Hertz vector potential	(2.87b)
Π_h	scalar magnetic potential	(3.35)
ϕ	spherical and cylindrical coordinates	(2.108)
$\phi_i(\boldsymbol{\rho})$	TM scalar transverse eigenfunfunction	(3.27)
Φ	electric scalar potential	(2.77)
$\bar{\phi}$	scalar function	(2.224)
$\bar{\psi}$	scalar function	(2.224)
$\varphi(\mathbf{r}, t)$	scalar field	(2.90)
$\bar{\varphi}$	scalar function	(3.76b)
$\Phi_m(t)$	flux of the magnetic induction	(2.9)
$p_v(\boldsymbol{r}, t)$	power loss density	(2.34)
Ψ	magnetic scalar potential	(2.84b)
$\psi_i(\boldsymbol{\rho})$	TE transverse scalar eigenfunction	(3.37a)
q	positive real function	(2.127)
$\widetilde{\mathbf{Q}}$	matrix	(5.34b)
$Q(t)$	charge	(2.8)
r	spherical coordinates: radial distance	(2.108)
$\boldsymbol{r}$	vector indicating a position in space	(2.1a)
$\Re$	real part of a complex quantity	(2.11)
ρ	cylindrical coordinates: radial distance	(2.108)
$\boldsymbol{\rho}$	transverse radial vector	Figure (3.1)
$\rho_e(\boldsymbol{r}, t)$	electric charge density	(2.1c)
$\rho_m(\boldsymbol{r})$	magnetic charge distributions,	(2.14d)
s	contour along surface S	Figure (3.1)
S	surface	(2.4)
$\bar{s}$	Laplace transform variable	(2.15)
$\boldsymbol{S}(\boldsymbol{r}, t)$	Poynting's vector	(2.36)
$S(u)$	sources	(2.169)
$\overline{S}(u_{1,2})$	sources impressed at the boundaries of the domain	(2.170)
S_α	expansion coefficients	(5.7)
$\mathbf{S}$	array	(5.9)
$\mathscr{S}'(\boldsymbol{r}, \boldsymbol{r}')$	TM Green's function	(3.77b)
$\mathscr{S}''(\boldsymbol{r}, \boldsymbol{r}')$	TE Green's function	(3.78b)
$\mathscr{S}_d'$	TM Green's function, inhomogeneous media	(3.91c)
$\mathscr{S}_d''$	TE Green's function, inhomogeneous media	(3.92e)

Symbol	Description	Reference
t	time variable	(2.1a)
$\boldsymbol{T}(\boldsymbol{r})$	complex Poynting's vector	(2.43)
τ	stands for either u, v or w	(2.94)
θ	spherical coordinates	(2.108)
$\mathbf{t}_s^\xi$	$\xi = \alpha, \beta$ array	(5.103)
$T^I(z, z')$	TL current modal Green's function	(3.61b)
$T^V(z, z')$	TL voltage modal Green's function	(3.61a)
$\hat{T}_e$	Transfer electic operator	(2.243b)
$\hat{T}_m$	impedance operator	(2.243b)
$\left(\hat{T}_k\right)_p$	transformation operators	(5.86)
$\mathscr{T}_e$	dyadic electric transfer function	(3.79)
$\mathscr{T}_m$	dyadic magnetic transfer function	(3.79)
U	complex function	(2.90)
u	scalar variable	(2.90)
$\boldsymbol{u}_n(\boldsymbol{r})$	vector basis functions	(5.60a)
$u_>$	scalar variable taking the value of u or u'	(2.180)
$u_<$	scalar variable taking the value of u or u'	(2.181)
$\boldsymbol{U}$	generic vector function	(2.4)
v	scalar variable	(2.90)
$\boldsymbol{v}_n(\boldsymbol{r})$	vector basis functions	(5.60b)
V	volume	(2.4)
$\boldsymbol{V}$	generic vector function	(2.4)
∂V	boundary of volume V	(2.38)
V_n^ξ	$\xi = \alpha, \beta$, field amplitudes of the electric fields	(5.22b)
V	array for the expansions coefficients of the electric field	(5.25)
V^ξ	$\xi = \alpha, \beta$, arrays for the expansions coefficients of the electric field	(5.23)
$V_i'(z)$	TM modal voltages	(3.31)
$V_i''(z)$	TE modal voltages	(3.40)
$v_i'(z)$	TM mode currents	(3.43d)
$v_i''(z)$	TE mode currents	(3.43d)
w	scalar variable	(2.90)
w	weight function, positive real function	(2.127)
w_α	α-th weight(test) function	(5.3)
$W(\bar{F}, F)$	Wronskian	(2.135)
$w_e(\boldsymbol{r}, t)$	electric energy density	(2.28)
$\overline{w}_e$	time averages of the electric energy density	(2.47)
$w_m(\boldsymbol{r}, t)$	magnetic energy density	(2.28)
$\overline{w}_m$	time averages of the magnetic energy density	(2.48)

Symbol	Description	Reference
x	cartesian coordinate	(2.98)
y	cartesian coordinate	(2.98)
Y	input admittance	(5.46b)
Y_i'	TM modal admittance	(3.34)
Y_i''	TE modal admittance	(3.42)
$Y(z,z')$	TL current modal Green's function	(3.61b)
$Y_{m,n}(\omega)$	admittance matrix term	(5.64b)
$\hat{Y}$	admittance operator	(2.243b)
$\mathscr{Y}$	dyadic admittance	(3.79)
$\mathscr{Y}^0$	dyadic impedance	(5.59)
$\mathscr{Y}^\lambda$	dyadic impedance	(5.59)
$\widetilde{\mathbf{Y}}_\lambda(\omega)$	multiport admittance matrix	(5.66)
z	cartesian and cylindrical coordinate	(2.98)
$\boldsymbol{z}_0$	unit vector in the z direction	(3.3)
Z	input impedance	(5.46a)
Z_0	characteristic impedance	(5.46a)
Z_i'	TM modal impedance	(3.34)
Z_i''	TE modal impedance	(3.42)
Z_{mn}^{+TM}	wave impedances for the outward propagating TM spherical modes	(5.79a)
Z_{mn}^{+TE}	wave impedances for the outward propagating TE spherical modes	(5.79b)
$Z(z,z')$	TL voltage modal Green's function	(3.61a)
$Z_{m,n}(\omega)$	impedance matrix term	(5.64a)
$\hat{Z}$	impedance operator	(2.243b)
$\mathscr{Z}$	dyadic impedance	(3.79)
$\mathscr{Z}^0$	dyadic impedance	(5.58)
$\mathscr{Z}^\lambda$	dyadic impedance	(5.58)
$\widetilde{\mathbf{Z}}_\lambda(\omega)$	multiport impedance matrix	(5.70)
ζ	variable	(4.31)
ζ_β	eigenvalues	(4.28)
$\bar{\zeta}$	free space admittance	(3.2)
$\langle \bar{F}, F \rangle$	inner product definition	(5.2)
$\frac{\partial}{\partial t}$	partial derivative with respect to t	(2.33)
$\nabla \cdot \boldsymbol{U}(\boldsymbol{r})$	divergence	(2.4)
$\nabla \times \boldsymbol{U}(\boldsymbol{r})$	curl	(2.5)
∇_t	transverse gradient operator	(3.3)

Index